AF863254

Changhua Bao

Electronic Band Structure Engineering and Ultrafast Dynamics of Dirac Semimetals

Doctoral Thesis accepted by
Tsinghua University, Beijing, China

清華大学出版社
TSINGHUA UNIVERSITY PRESS

Springer

Author
Dr. Changhua Bao
Department of Physics
Tsinghua University
Beijing, China

Supervisor
Prof. Zhou Shuyun
Tsinghua University
Beijing, China

ISSN 2190-5053 ISSN 2190-5061 (electronic)
Springer Theses
ISBN 978-981-99-5324-0 ISBN 978-981-99-5325-7 (eBook)
https://doi.org/10.1007/978-981-99-5325-7

Jointly published with Tsinghua University Press
The print edition is not for sale in China (Mainland). Customers from China (Mainland) please order the print book from: Tsinghua University Press.
ISBN of the Co-Publisher's edition: 978-7-302-63906-0.

© Tsinghua University Press 2023

This work is subject to copyright. All rights are solely and exclusively licensed by the Publisher, whether the whole or part of the material is concerned, specifically the rights of reprinting, reuse of illustrations, recitation, broadcasting, reproduction on microfilms or in any other physical way, and transmission or information storage and retrieval, electronic adaptation, computer software, or by similar or dissimilar methodology now known or hereafter developed.
The use of general descriptive names, registered names, trademarks, service marks, etc. in this publication does not imply, even in the absence of a specific statement, that such names are exempt from the relevant protective laws and regulations and therefore free for general use.
The publishers, the authors, and the editors are safe to assume that the advice and information in this book are believed to be true and accurate at the date of publication. Neither the publishers nor the authors or the editors give a warranty, expressed or implied, with respect to the material contained herein or for any errors or omissions that may have been made. The publishers remain neutral with regard to jurisdictional claims in published maps and institutional affiliations.

This Springer imprint is published by the registered company Springer Nature Singapore Pte Ltd.
The registered company address is: 152 Beach Road, #21-01/04 Gateway East, Singapore 189721, Singapore

Paper in this product is recyclable.

Supervisor's Foreword

Dirac semimetals exhibit unique band structures and novel light-matter interactions. Revealing their band structures and ultrafast dynamics is therefore important. Dr. Changhua Bao's Ph.D. thesis focuses on the instrumentation development of cutting-edge time- and angle-resolved photoemission spectroscopy (TrARPES) for revealing the dynamics of Dirac semimetals, as well as scientific progress on the band engineering and ultrafast dynamics of two typical Dirac semimetals enabled by the newly developed instrumentation. The TrARPES system that he developed covers a uniquely tunable probe photon energy range of 5.3~7.0 eV, with high energy, momentum, and time resolutions, thus providing important opportunities for investigating the ultrafast dynamics of 3D quantum materials. By using this unique system, he reported the experimental evidence of chiral symmetry breaking and flat band in Kekulé-ordered graphene for the first time. In addition, thanks to the widely tunable probe photon energy of the TrARPES system, he reported the ultrafast dynamics of 3D Dirac fermion in a typical 3D Dirac semimetal Cd_3As_2 for the first time. A long-lived population inversion effect is revealed, which paves an important step toward exploring exotic states such as the dynamical excitonic insulating state. This book advances the TrARPES instrumentation as well as an understanding of the electronic structure and dynamics of Dirac semimetals.

Beijing, China
September 2023

Prof. Zhou Shuyun

Abstract

Dirac semimetals exhibit unique electronic band structure, novel physical properties with rich light-matter interaction, and potential applications. Revealing the band engineering and ultrafast dynamics of Dirac semimetals is therefore important. In this thesis, we develop the first time- and angle-resolved photoemission spectroscopy (TrARPES) with tunable probe photon energy and investigate the band structure engineering and ultrafast dynamics of Dirac semimetals by utilizing such a unique TrARPES system. Below are major scientific achievements:

(1) We report the development of a TrARPES system with a highly tunable probe photon energy from 5.3 to 7.0 eV. Our TrARPES system has high energy, momentum, and time resolution (280–320 fs), which opens up new opportunities for exploring ultrafast dynamics in 3D quantum materials.
(2) We realize the long-sought-after chiral symmetry breaking (CSB) via Li intercalation of graphene and provide direct experimental evidences. The CSB is evidenced by the gap opening at the Dirac point, Kekeulé-O type modulation, and chirality mixing near the gap edge. In addition, we also report the experimental observation of an extended flat band below E_F in this system with strong electron-phonon interaction which coexists and codevelops with the Kekulé order, thus providing opportunities for investigating CSB-related physics, flat band-related instabilities, and its interplay with the Kekulé order.
(3) We reveal the ultrafast dynamics of Dirac fermions in a model 3D Dirac semimetal Cd_3As_2 by using our TrARPES system with a tunable probe photon energy. The energy and momentum-resolved relaxation rate shows a linear dependence on the energy, suggesting Dirac fermion cooling through intraband relaxation. Moreover, a population inversion with a long lifetime of 3.0 ps is reported, which paves an important step toward exploring the possible dynamic exotic excitonic insulator.

This thesis advances the research of the electronic structure of Dirac semimetal from two aspects. First, it identifies the Kekulé-ordered graphene as a new system for exploring chiral symmetry breaking-related physics including electronic fractionalization and topology effect and flat band-induced instability, including chiral superconductivity, especially providing a rare system to investigating their interplay. Secondly, this thesis solves the long-standing challenge of directly visualizing the non-equilibrium band structure of 3D Dirac semimetal and opens up new opportunities for exploring the light-matter interaction in 3D quantum materials, especially the light-induced topological phase transitions in 3D topological material through Floquet engineering.

Keywords Dirac semimetal · Band engineering · Ultrafast dynamics · Angle-resolved photoemission spectroscopy (ARPES) · Time- and angle-resolved photoemission spectroscopy (TrARPES)

Parts of this thesis have been published in the following documents
Journal Publications

Changhua Bao†, Hongyun Zhang†, Teng Zhang, *et al.* Experimental evidence of chiral symmetry breaking in a Kekulé-ordered graphene. Phys. Rev. Lett., 2021; 126:206804.

Changhua Bao, Peizhe Tang*, Dong Sun* and Shuyun Zhou* Light-induced emergent phenomena in 2D materials and topological materials. Nat. Rev. Phys., 2022; 4:33.

Changhua Bao†, Hongyun Zhang†, Xi Wu, *et al.* Coexistence of extended flat band and Kekulé order in Li-intercalated graphene. Phys. Rev. B, 2022; 105:L161106.

Changhua Bao, Qian Li, Sheng Xu, *et al.* Population inversion and Dirac fermion cooling in 3D Dirac semimetal Cd_3As_2. Nano Lett., 2022; 22:1138.

Changhua Bao, Haoyuan Zhong, Shaohua Zhou, *et al.* Ultrafast time- and angle-resolved photoemission spectroscopy with widely tunable probe photon energy of 5.3–7.0 eV for investigating dynamics of three-dimensional materials. Rev. Sci. Instrum., 2022; 93:013902.

Changhua Bao, Laipeng Luo, Hongyun Zhang, *et al.* Full diagnostics and optimization of time resolution for time- and angle-resolved photoemission spectroscopy. Rev. Sci. Instrum., 2021; 92:033904.

Contents

1 Introduction 1
1.1 Overview of Dirac Semimetal 1
1.1.1 2D Dirac Semimetal 1
1.1.2 3D Dirac Semimetal 3
1.2 Intriguing Light-Matter Interaction in Dirac Semimetal 4
1.2.1 Ultrafast Dynamics and Population Inversion 4
1.2.2 Floquet Engineering in Dirac Semimetal 6
References 10

2 Experimental Techniques 15
2.1 Angle-Resolved Photoemission Spectroscopy 15
2.1.1 Basic Principles 15
2.1.2 Instruments 18
2.2 Time- and Angle-Resolved Photoemission Spectroscopy 20
2.2.1 Basic Principles 21
2.2.2 Special Technical Considerations 22
References 24

3 Development of Novel TrARPES with Tunable Probe Photon Energy for 3D Quantum Materials 27
3.1 Research Motivation 27
3.2 TrARPES with Tunable Probe Photon Energy 28
3.2.1 Experimental Setup 28
3.2.2 Tunable Probe Photon Energy 30
3.2.3 Considerations for the Time Resolution 33
3.2.4 TrARPES Functionality 39
3.3 Conclusions 40
References 40

4 Chiral Symmetry Breaking in Kekulé-Ordered Graphene 43
4.1 Research Motivation 43
4.2 Experimental Evidence of Chiral Symmetry Breaking 44
4.2.1 Preparation of Kekulé-Ordered Graphene 44
4.2.2 Evidence of CSB 1: Band Gap Opening 48
4.2.3 Evidence of CSB 2: Kekulé-O Order 50
4.2.4 Evidence of CSB 3: Chirality Mixing 51
4.3 Conclusion 55
References 55

5 Coexistence of Flat Band and Kekulé Order 59
5.1 Research Motivation 59
5.2 Extremely High Doping Level 60
5.3 Strong el-ph Coupling 62
5.4 Co-development of Flat Band and Kekulé Order 63
5.5 Conclusion 65
References 66

6 Dirac Fermion Cooling in 3D Dirac Semimetal Cd_3As_2 69
6.1 Research Motivation 69
6.2 Revealing the 3D Dirac Cone 70
6.3 Photoexcited Carrier Dynamics 71
6.4 Ultrafast Thermodynamics 73
6.5 Population Inversion 73
6.6 Conclusion 76
References 76

7 Summary and Prospects 79
References 80

Symbols and Acronyms

2D	Two dimensional
3D	Three dimensional
AFM	Atomic force microscope
ARPES	Angle-resolved photoemission spectroscopy
BBO	β-BaB_2O_4
BS	Beam splitter
BZ	Brillouin zone
CB	Conduction band
CCD	Charge-coupled device
CPL	Circular polarized light
CSB	Chiral symmetry breaking
DFG	Difference frequency generation
DFT	Density functional theory
DOS	Density of states
DS	Delay stage
E	Energy
EDCs	Energy distribution curves
EPC	Electron-phonon coupling
eV	Electron volt
FB	Fundamental beam
FH	Fourth harmonic
FHG	Fourth harmonic generation
FROG	Frequency-resolved optical gating
fs	Femtosecond
FWHM	Full width at half maximum
k	Wave vector
KBBF	$KBe_2BO_3F_2$
KBBF-PCT	$KBe_2BO_3F_2$ crystal with prism-coupled technique
LEED	Low energy electron diffraction
MABLG	Magic-angle twisted bilayer graphene
MDCs	Momentum distribution curves

meV	Milli electron Volt
ML	Monolayer
NanoARPES	Nanospot angle-resolved photoemission spectroscopy
ps	Picosecond
QED	Quantum electrodynamics
SFG	Sum-frequency generation
SH	Second harmonic
SHG	Second harmonic generation
SOC	Spin-orbital coupling
STM	Scanning tunneling microscopy
TI	Topological insulator
TrARPES	Time- and angle-resolved photoemission spectroscopy
TTM	Two-temperature model
UHV	Ultra-high vacuum
UV	Ultraviolet
VB	Valence band
VHS	Van Hove singularity
VUV	Vacuum ultraviolet
WF	Weyl Fermion

Chapter 1
Introduction

1.1 Overview of Dirac Semimetal

Dirac semimetal [1, 2] shows unique linear electronic dispersion and point-like Dirac node as shown in Fig. 1.1b, in analogy to the massless Dirac fermions in particle physics [3], which is distinct from the typical parabolic dispersion in semiconductor as shown in Fig. 1.1b. It is an analog of quantum electrodynamics (QED) [4, 5], leading to a condensed-matter physics system to mimick quantum relativistic phenomena in high-energy physics and even the exotic physics which has no counterpart in high-energy physics. Such unique electronic structure also leads to promising properties for applications such as optoelectronics [6, 7].

1.1.1 2D Dirac Semimetal

A kind of two-dimensional (2D) materials represented by graphene show linear electronic dispersion and is called 2D Dirac semimetal, including silicene, germanene, graphynes, transition-metal oxides and so on [8]. Graphene is the first discovered 2D Dirac semimetal [9–12]. It is a two-dimensional sheet of carbon atoms arranged in hexagonal honeycomb lattice as shown in Fig. 1.2a, which is first fabricated by direct exfoliation [9, 10] and shows exceptionally excellent qualities [11]. For decades, the 2D crystal has been thought to be unstable [13, 14], the discovery of graphene directly leads to the explosion of the new area of two-dimensional materials [15]. The unique atomic structure of graphene also leads to an exotic electronic structure: six linear Dirac cones at momentum valleys. Such exotic electronic structure creates a new playground for intriguing physics, such as quantum Hall effect [9, 10].

There exist two atoms A and B in one unit cell of graphene, which is called sublattice freedom. This can be regarded as a two-level system to describe the electron occupies in A or B sublattice, which can be used to define the pseudospin [16] in

© Tsinghua University Press 2023
C. Bao, *Electronic Band Structure Engineering and Ultrafast Dynamics of Dirac Semimetals*, Springer Theses, https://doi.org/10.1007/978-981-99-5325-7_1

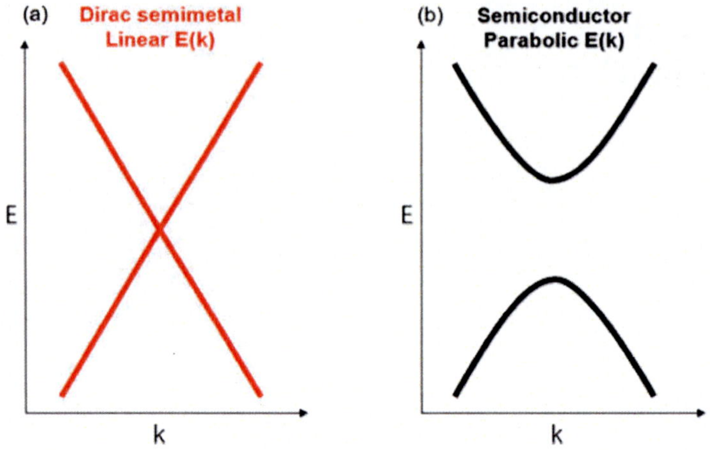

Fig. 1.1 Linear dispersions in Dirac semimetal versus parabolic dispersions in the semiconductor

Fig. 1.2 2D Dirac semimetal graphene

analogy to the spin up and down. Massless Dirac fermions possess chirality defined by the projection of pseudospin on the momentum, and more interestingly, the Dirac fermions at different valleys show opposite chiralities. In particle physics, by coupling massless Dirac fermions with opposite chiralities, the chiral symmetry can be broken, thus leading to the mass generation [17] which is one of the cornerstones of particle physics. Graphene provides an ideal system to simulate chiral symmetry breaking (CSB) in a condensed matter physics system. Chiral symmetry breaking can be realized through intervalley coupling [18–23] induced by the Kekulé bond texture and the Dirac point can be gapped leading to massive Dirac fermions.

In addition to the physics near the Dirac point, the band structure at higher binding energy is also interesting. In pristine graphene, the Fermi level is around the Dirac point leading to a very small density of states at the Fermi level, in contrast, by tuning the doping level to 1/4 filling (1/4 electrons per unit cell), a van Hove singularity appears at the Fermi level leading to a very high density of states. Such a high density of states at the Fermi level can enhance the many-body interaction, which has attracted extensive interests with intriguing predictions such as instabilities toward d-wave superconductivity by repulsive electron-electron interaction [24–26], topological superconductivity [27], and chiral spin density wave [28, 29].

1.1.2 3D Dirac Semimetal

In addition to 2D Dirac semimetal, linear electronic dispersion can also exist in three dimension as shown in Fig. 1.3, which is described by the Hamiltonian,

$$H(\boldsymbol{k}) = \pm\hbar v \boldsymbol{k} \cdot \boldsymbol{\sigma} \tag{1.1}$$

where $\boldsymbol{k}$ is the wave vector in three-dimensional momentum space and $\boldsymbol{\sigma}$ is the Pauli matrice of spin. The material with such a Hamiltonian is called three-dimensional (3D) Weyl semimetal [30–39], which can be seen as a 3D analogue of graphene. The Hamiltonian is the same as the operator of chirality (the projection of spin on momentum), so the massless Weyl fermions possess the right-handed or left-handed chiralities corresponding to the $\pm$ in the Hamiltonian. In addition, the Weyl semimetal is topologically nontrivial. The Weyl point is the monopole of Berry curvature and the flux of Berry curvature through any surface enclosing the Weyl point is equal to the chirality. Therefore, the Weyl points always appear in pairs with opposite chiralities. Due to the nontrivial topology, novel topological surface states with open boundaries show up by connecting the Weyl points with opposite chiralities, so-called Fermi arc. The existence of Weyl points needs either inversion symmetry or time-reversal symmetry breaking. If the symmetry is reserved, two Weyl points with opposite chiralities merge together to form the 3D Dirac semimetal [40–42],

$$H(\boldsymbol{k}) = \begin{pmatrix} \hbar v \boldsymbol{k} \cdot \boldsymbol{\sigma} & 0 \\ 0 & -\hbar v \boldsymbol{k} \cdot \boldsymbol{\sigma} \end{pmatrix} \tag{1.2}$$

Because the Dirac point is consist of two Weyl points with opposite chiralities, the net Chern number is zero and not topologically protected. Instead, the Dirac point is protected by symmetry and robust even with spin-orbital coupling (SOC) in contrast to 2D Dirac semimetal graphene where the Dirac point can be gapped by SOC. In addition, the Fermi arcs can also appear in Dirac semimetal but are more subtle, because of the zero Chern number.

3D Dirac semimetal is first reported in Na_3Bi [41, 43, 44]. Two Dirac points along the [001] direction, Fermi arcs and linear dispersions along all three momentum directions are confirmed by ARPES measurements [43, 44]. However, Na_3Bi is not stable in the air which makes it difficult for further investigation and applications. In contrast, Cd_3As_2 is found to be a stable 3D Dirac semimetal [42, 45–48]. Two Dirac points along the [111] direction and linear dispersions along all three momentum directions are confirmed by ARPES measurements [45–47].

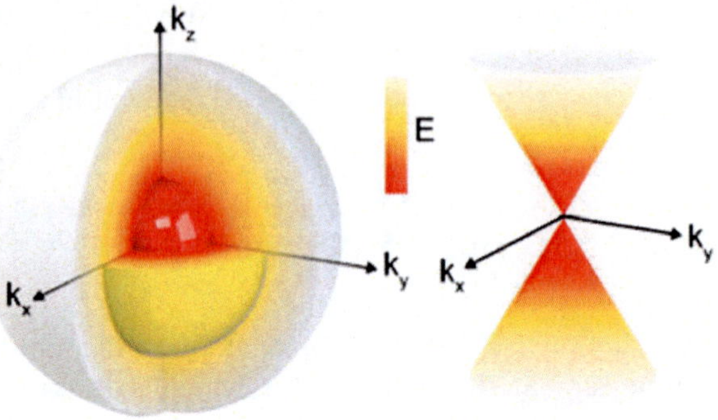

Fig. 1.3 3D Dirac cone and its projection on 2D plane

The topological properties of Dirac (or Weyl) semimetal strongly affect the materials' properties, leading to exotic transport behavior such as quantum anomalous Hall effect [31–33], the negative longitudinal magnetoresistance [49–52] and thermoelectric response [53–55]. In addition, the Dirac (or Weyl) semimetal also shows high mobility [48, 56, 57] connected with the massless feature and high Fermi velocity and has attracted extensive research interests [2, 58–61].

1.2 Intriguing Light-Matter Interaction in Dirac Semimetal

The unique electronic structure of Dirac semimetal, for example, linear dispersion, the vanished density of states at Dirac point, Berry curvature, and strong dependence on symmetries, leads to intriguing light-matter interaction which is distinct from the conventional insulator or metals. Studying the light-matter interaction in Dirac semimetal is important for both practical device application and fundamental physics.

1.2.1 Ultrafast Dynamics and Population Inversion

The unusual electronic structure of Dirac semimetal adds an extra restriction to the ultrafast dynamics of photoexcited carriers, leading to exotic relaxations of photocarriers. Before photoexcitation, the electrons stay in the lowest energy, for example, the valence band of the Dirac cone, and leaving the most of conduction band unoccupied. Upon photoexcitation, the electrons are excited from the valence band (VB) to the conduction band (CB) firstly due to the direct adsorption of photons. Due to the strong electron-electron scattering, the photoelectrons would relax to quasi-equilibrium states in tens to hundreds of fs. All the electrons balance with each other and reach the same temperature (electronic temperature) [62, 63]. However, the electronic temperature is still higher than the lattice, so the next step would be the energy transfer from electrons to phonons on a slower time scale. Finally, the whole system recoveries to the equilibrium states.

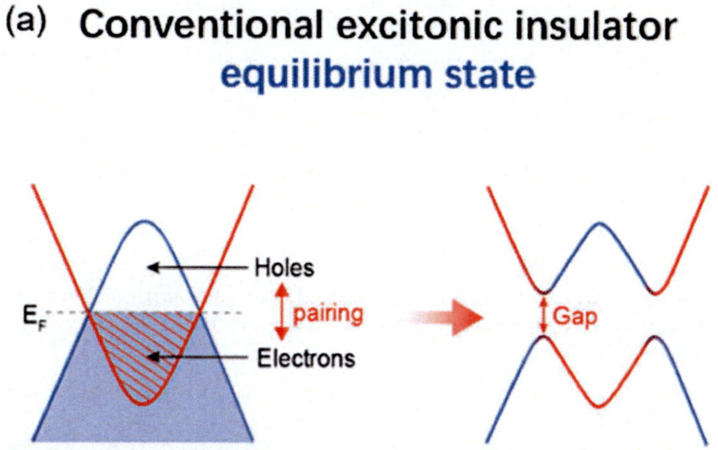

Fig. 1.4 Conventional excitonic insulator and dynamical excitonic insulator. **a** Conventional excitonic insulator in the equilibrium state

The above basic mechanism works for most materials including conventional metal and insulators. In Dirac semimetal, an important question is the interplay between electrons in the conduction band and holes in the valence band: whether they relax to phonons acting like one or two separated systems. That is to say which process is slower for interband and intraband relaxations. The Dirac semimetal has a unique Dirac point near the Fermi level which suppresses the phase space of the interband relaxation as a bottleneck, so slow interband relaxation and long-lived electrons and holes are expected [64–68], so-called population inversion. Such population inversion is potentially useful for photoelectronics and laser applications especially in THz and midinfrared range due to the gapless electronic structure [7, 69, 70].

In addition, the population inversion also provides new opportunities for realizing new phases of matter which has no counterpart in equilibrium states, such as the dynamical excitonic insulator [71]. In a conventional excitonic insulator, the intrinsic electrons and holes pair with each other due to the strong electron-hole interaction [72–77]. A band gap is opened at the Fermi energy and the system becomes an insulator from metal as shown in Fig. 1.4a. Actually, the electrons and holes can also be generated by photoexcitation. The strong electron-hole interaction also leads to the formation of excitonic insulator but only on an ultrafast time scale, so-called dynamical excitonic insulator [78, 79].

For both potential applications for laser and dynamical excitonic insulators, long-lived population inversion is highly desired. Experimentally, the population inversion has been reported in 2D Dirac semimetal graphene by time- and angle-resolved photoemission spectroscopy (TrARPES) study [65]. The photoexcited electrons in the conduction band above Fermi energy are clearly observed and it relaxes very fast, and the electrons show distinct chemical potential from the holes, which means the formation of population inversion. However, the electron and hole balance quickly with each other and share the same chemical potential, suggesting the lifetime of pop-

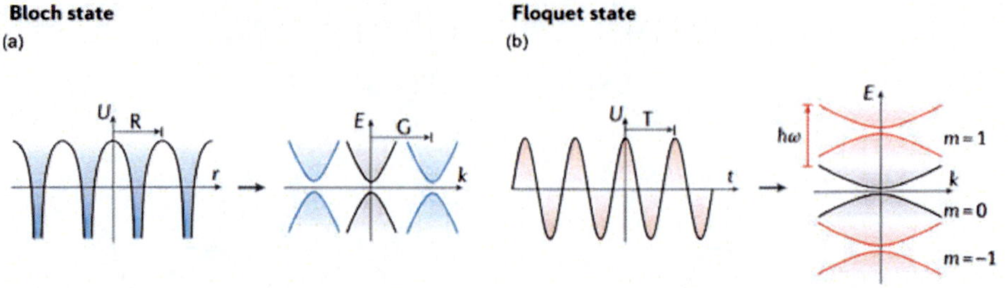

Fig. 1.5 Floquet states in time-periodic electric field. **a** Spatially-periodic potential and Bloch bands in the k-space. **b** Time-periodic potential and Floquet bands in energy. Reproduced, with permission from [91]

ulation inversion is only 130 fs, which is dominated by the fast Auger recombination. Because even the phase space is zero but the Auger interaction strength diverges, leading to a finite Auger recombination rate in 2D Dirac semimetal [80–82].

The Auger recombination can be suppressed in 3D Dirac semimetal, it provides more promising opportunities for exploring long-lived population inversion. In 3D Dirac semimetal, the phase space is still zero but the Auger interaction strength becomes finite, leading to suppressed Auger recombination. Long-lived population inversion up to hundreds of microseconds has been predicted [68].

1.2.2 Floquet Engineering in Dirac Semimetal

Another interesting light-matter interaction in Dirac semimetal is the Floquet engineering. Light can be regarded as a time-periodic electric field and can couple with electrons in solid-state materials, leading to the formation of quasi-energy sidebands through multi-photon absorption or emission, which are called the Floquet states [83–91]. In solid state physics, the spatially periodic Hamiltonian results in Bloch states whose energy bands have the discrete translational symmetry in the momentum space as shown in Fig. 1.5a. In analogy, by applying a time-periodic driving field to the solid-state system, it is possible to form a steady Floquet state and the related Hamiltonian has the discrete time-translational symmetry as shown in Fig. 1.5b. For Bloch states, hybridization with the folded bands at the BZ boundary (in momentum) can lead to a band gap opening. Similarly, in the Floquet theory, the electronic properties of the host materials could be changed by coupling with the optical field. For example, a gap can open for 2D Dirac materials at the band crossing and the first Floquet BZ boundary. This application of a time-periodic driving field can be used as a control knob for engineering non-equilibrium material properties, dubbed as Floquet engineering.

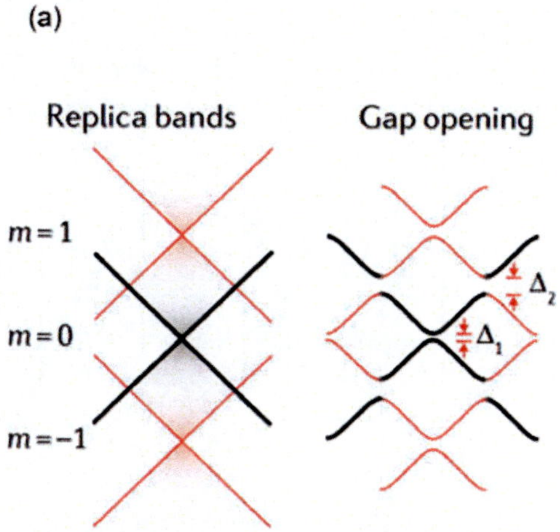

Fig. 1.6 Floquet engineering in 2D Dirac system. **a** Floquet engineering in 2D Dirac system leading to Floquet sidebands (red) and resonant gap opening at the crossing point. Reproduced, with permission from [91]

1.2.2.1 Floquet Engineering in 2D Dirac Semimetal

Floquet-Bloch states were theoretically proposed in 2D Dirac semimetal graphene, where the application of a circular polarized light (CPL) is predicted to lead to a gap opening at the Dirac point and a photo-induced Hall current [92] as shown in Fig. 1.6. Unlike the conventional Hall effect which is induced by a static magnetic field in equilibrium, the photo-induced Hall current is caused by strong light-matter interaction which breaks the time-reversal symmetry. The topological properties of such non-equilibrium, periodically driven systems are further analyzed [93], and an analogy to the Haldane model [5] is discussed. A nonzero Chern number has been obtained, and topologically protected edge states and quantum Hall effect have been proposed [94–99]. Floquet engineering has also been extended to other Dirac fermion systems, such as the surface state of topological insulators [94].

Experimentally, Floquet sidebands of the topological surface states have been observed in TI Bi_2Se_3 in the ultrafast time scale when the pump overlaps coherently with the probe pulse [100, 101] as shown in Fig. 1.7. There are two experimental signatures to indicate Floquet engineering. One is the observation of a dynamical gap generation of 62 meV at the crossing points between the topological surface state and sideband replicas along the momentum direction perpendicular to the electric field for both linear (p and s polarizations) and CPL. The other is the observation of a gap of 53 meV at the Dirac point of the surface state due to the breaking of time-reversal symmetry by CPL. In addition, by switching the pump polarization between p and s, selective excitations of Floquet states versus Volkov states have been demonstrated, which correspond to photon-dressed initial electronic states (inside the sample) and photon-dressed free-electron-like final states (after photo-excited from the sample into the vacuum) respectively.

Although graphene has been proposed as a model system for realizing topological Floquet states since 2009 [92], signatures of Floquet states in graphene have

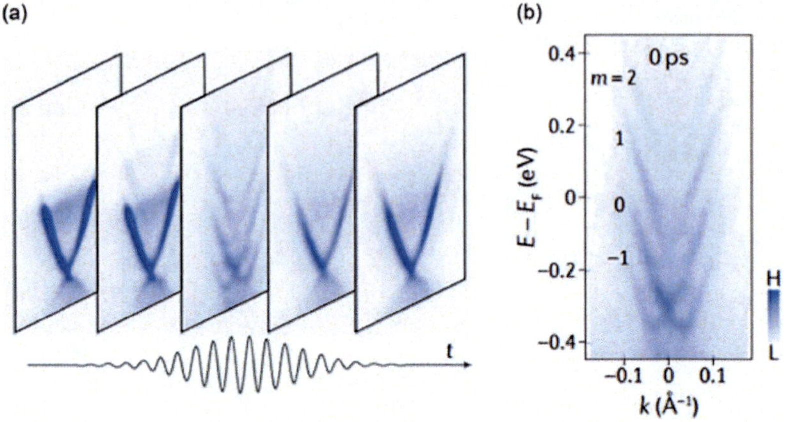

Fig. 1.7 Experimental realization of Floquet engineering in 2D Dirac surface states of topological insulator (TI). **a** TrARPES spectrum of surface Dirac cone at different delay times. **b** TrARPES spectrum at the zero delay time. Reproduced, with permission from [91]

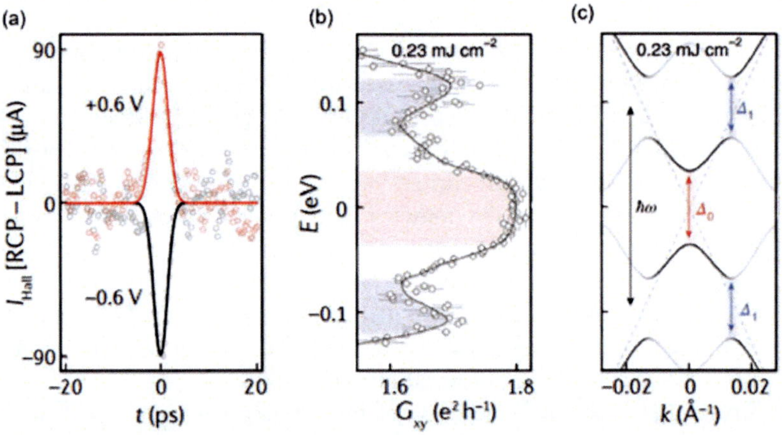

Fig. 1.8 Experimental realization of Floquet engineering in graphene. **a** Light-induced anomalous Hall current signals. **b** Light-induced Hall conductance as a function of energy. **c** Effective band structure under light excitation using the Floquet theory. Reproduced, with permission from [91]

been demonstrated only recently by light-induced anomalous Hall effect (AHE) or photovoltaic Hall effect [102]. CPL excitation at a photon energy of 191 meV is used to generate a dynamical gap in graphene, and anomalous Hall currents are reported to be possible characteristics of Floquet states according to the following two experimental features. First, the anomalous Hall current is transverse to the applied longitudinal current and it reverses polarity for opposite light helicities as shown in Fig. 1.8a. Second, a conductance plateau is observed with a width of 60 meV when the Fermi energy is tuned inside the gap as shown in Fig. 1.8b, suggesting a gap opening consistent with the calculated light-induced gap of 69 meV (Fig. 1.8c).

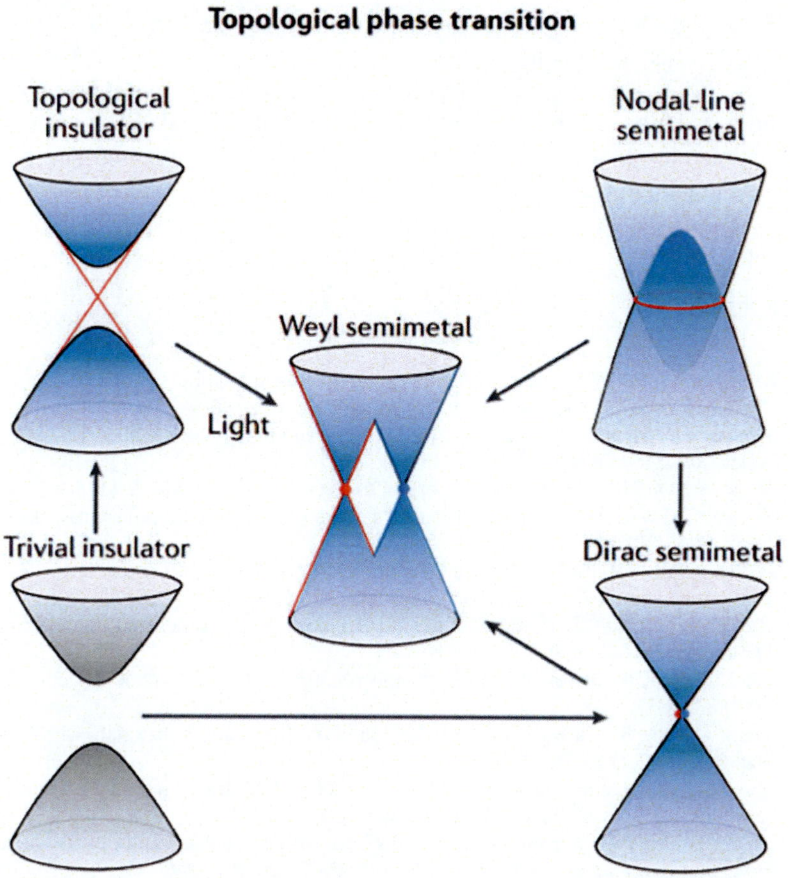

Fig. 1.9 A schematic summary of light-induced topological phase transitions in 3D topological materials. Reproduced, with permission from [91]

The gap opening in graphene inferred by time-resolved transport measurements of such Floquet engineering could in principle be detectable by TrARPES, which still awaits experimental progress.

1.2.2.2 Floquet Engineering in 3D Dirac Semimetal

Floquet engineering can also be applied to 3D Dirac/Weyl semimetal to achieve highly tunable topological properties in non-equilibrium as shown in Fig. 1.9. Floquet engineering can be used to dynamically generate a Weyl semimetal from a ground state TI by renormalizing the bulk energy gap of a TI [103]. For Dirac semimetal as the ground state, CPL could break the time-reversal symmetry and split Dirac fermions into a pair of Weyl fermions (WFs) [104–106], and their separation is tunable by light intensity or direction and the WF can be even tilted to type-II WF with highly tilted Weyl cones.

In this thesis, we will study both the equilibrium and non-equilibrium electronic structures of Dirac semimetal by utilizing ARPES and TrARPES for exploring the chiral symmetry breaking and 1/4 doping in 2D Dirac semimetal and population inversion in 3D Dirac semimetal.

References

1. Murakami S (2007) Phase transition between the quantum spin Hall and insulator phases in 3D: emergence of a topological gapless phase. New J Phys 9(9):356
2. Armitage N, Mele E, Vishwanath A (2018) Weyl and Dirac semimetals in three-dimensional solids. Rev Mod Phys 90(1):015001
3. Dirac PAM (1928) The quantum theory of the electron. Proc R Soc A 117(778):610–624
4. Semenoff GW (1984) Condensed-matter simulation of a three-dimensional anomaly. Phys Rev Lett 53(26):2449
5. Haldane FDM (1988) Model for a Quantum Hall effect without Landau levels: Condensed-matter realization of the "parity anomaly. Phys Rev Lett 61:2015–2018
6. Bonaccorso F, Sun Z, Hasan T et al (2010) Graphene photonics and optoelectronics. Nat Photonics 4(9):611–622
7. Liu J, Xia F, Xiao D et al (2020) Semimetals for high-performance photodetection. Nat Mater 19(8):830–837
8. Wang J, Deng S, Liu Z et al (2015) The rare two-dimensional materials with Dirac cones. Natl Sci Rev 2(1):22–39
9. Novoselov KS, Geim AK, Morozov SV et al (2005) Two-dimensional gas of massless Dirac fermions in graphene. Nature 438(7065):197–200
10. Zhang Y, Tan YW, Stormer HL et al (2005) Experimental observation of the quantum Hall effect and Berry's phase in graphene. Nature 438(7065):201–204
11. Geim AK, Novoselov KS (2007) The rise of graphene. Nat Mater 6:183–191
12. Castro Neto AH, Guinea F, Peres NMR et al (2009) The electronic properties of graphene. Rev Mod Phys 81:109–162
13. Peierls R (1935) Quelques propriétés typiques des corps solides. In: Annales de I. H. Poincare, vol 5, pp 177–222
14. Landau LD (1937) Zur theorie der phasenumwandlungen II. Phys Z Sowjetunion 11(545):26–35
15. Geim AK, Grigorieva IV (2013) Van der Waals heterostructures. Nature 499(7459):419–425
16. Pesin D, MacDonald AH (2012) Spintronics and pseudospintronics in graphene and topological insulators. Nat Mater 11(5):409–416
17. Nambu Y (1960) Quasi-particles and gauge invariance in the theory of superconductivity. Phys Rev 117:648–663
18. Chamon C (2000) Solitons in carbon nanotubes. Phys Rev B 62:2806–2812
19. Cheianov VV, Fal'ko VI, Syljuåsen O et al (2009) Hidden Kekulé ordering of adatoms on graphene. Solid State Commun 149(37):1499–1501
20. González J (2010) Renormalization group approach to chiral symmetry breaking in graphene. Phys Rev. B 82:155404
21. Lee SH, Chung HJ, Heo J et al (2011) Band gap opening by two-dimensional manifestation of peierls instability in graphene. ACS Nano 5(4):2964–2969
22. Semenoff GW (2012) Chiral symmetry breaking in graphene. Phys Scr 146:014016
23. Roy B, Kennett MP, Das Sarma S (2014) Chiral symmetry breaking and the quantum Hall effect in monolayer graphene. Phys Rev B 90:201409

24. Black-Schaffer AM, Doniach S (2007) Resonating valence bonds and mean-field d-wave superconductivity in graphite. Phys Rev B 75:134512
25. Raghu S, Kivelson SA, Scalapino DJ (2010) Superconductivity in the repulsive Hubbard model: An asymptotically exact weak-coupling solution. Phys Rev B 81:224505
26. Pathak S, Shenoy VB, Baskaran G (2010) Possible high-temperature superconductivity with a $d + id$ pairing symmetry in doped graphene. Phys Rev B 81:085431
27. Nandkishore R, Levitov LS, Chubukov AV (2012) Chiral superconductivity from repulsive interactions in doped graphene. Nat Phys 8:158
28. Wang WS, Xiang YY, Wang QH et al (2012) Functional renormalization group and variational Monte Carlo studies of the electronic instabilities in graphene near $\frac{1}{4}$ doping. Phys Rev B 85:035414
29. Jiang S, Mesaros A, Ran Y (2014) Chiral spin-density wave, spin-charge-chern liquid, and $d + id$ superconductivity in 1/4-doped correlated electronic systems on the honeycomb lattice. Phys Rev X 4:031040
30. Wan X, Turner AM, Vishwanath A et al (2011) Topological semimetal and Fermi-arc surface states in the electronic structure of pyrochlore iridates. Phys Rev B 83:205101
31. Yang KY, Lu YM, Ran Y (2011) Quantum Hall effects in a Weyl semimetal: Possible application in pyrochlore iridates. Phys Rev B 84(7):075129
32. Burkov AA, Balents L (2011) Weyl semimetal in a topological insulator multilayer. Phys Rev Lett 107:127205
33. Xu G, Weng H, Wang Z et al (2011) Chern semimetal and the quantized anomalous Hall effect in $HgCr_2Se_4$. Phys Rev Lett 107(18):186806
34. Witczak-Krempa W, Kim YB (2012) Topological and magnetic phases of interacting electrons in the pyrochlore iridates. Phys Rev B 85(4):045124
35. Weng H, Fang C, Fang Z et al (2015) Weyl semimetal phase in noncentrosymmetric transition-metal monophosphides. Phys Rev X 5(1):011029
36. Huang SM, Xu SY, Belopolski I et al (2015) A Weyl fermion semimetal with surface fermi arcs in the transition metal monopnictide TaAs class. Nat Commun 6(1):1–6
37. Xu SY, Belopolski I, Alidoust N et al (2015) Discovery of a Weyl fermion semimetal and topological Fermi arcs. Science 349(6248):613–617
38. Lv BQ, Weng HM, Fu BB et al (2015) Experimental discovery of Weyl semimetal TaAs. Phys Rev X 5:031013
39. Lu L, Wang Z, Ye D et al (2015) Experimental observation of Weyl points. Science 349(6248):622–624
40. Young SM, Zaheer S, Teo JCY et al (2012) Dirac semimetal in three dimensions. Phys Rev Lett 108:140405
41. Wang Z, Sun Y, Chen XQ et al (2012) Dirac semimetal and topological phase transitions in A_3Bi (A= Na, K, Rb). Phys Rev B 85:195320
42. Wang Z, Weng H, Wu Q et al (2013) Three-dimensional Dirac semimetal and quantum transport in Cd_3As_2. Phys Rev B 88:125427
43. Liu Z, Zhou B, Zhang Y et al (2014) Discovery of a three-dimensional topological Dirac semimetal, Na_3Bi. Science 343(6173):864–867
44. Xu SY, Liu C, Kushwaha SK et al (2015) Observation of Fermi arc surface states in a topological metal. Science 347(6219):294–298
45. Liu ZK, Jiang J, Zhou B et al (2014) A stable three-dimensional topological Dirac semimetal Cd_3As_2. Nat Mater 13(7):677–81
46. Neupane M, Xu SY, Sankar R et al (2014) Observation of a three-dimensional topological Dirac semimetal phase in high-mobility Cd_3As_2. Nat Commun 5:3786
47. Ali MN, Gibson Q, Jeon S et al (2014) The crystal and electronic structures of Cd_3As_2, the three-dimensional electronic analogue of graphene. Inorg Chem 53(8):4062–4067
48. Liang T, Gibson Q, Ali MN et al (2015) Ultrahigh mobility and giant magnetoresistance in the Dirac semimetal Cd_3As_2. Nat Mater 14(3):280
49. Son D, Spivak B (2013) Chiral anomaly and classical negative magnetoresistance of Weyl metals. Phys Rev B 88(10):104412

50. Burkov A (2015) Negative longitudinal magnetoresistance in Dirac and Weyl metals. Phys Rev B 91(24):245157
51. Xiong J, Kushwaha SK, Liang T et al (2015) Evidence for the chiral anomaly in the Dirac semimetal Na_3Bi. Science 350(6259):413–416
52. Burkov AA, Kim YB (2016) Z2 and chiral anomalies in topological Dirac semimetals. Phys Rev Lett 117(13):136602
53. Hirschberger M, Kushwaha S, Wang Z et al (2016) The chiral anomaly and thermopower of Weyl fermions in the half-Heusler GdPtBi. Nat Mater 15(11):1161–1165
54. Chen Q, Fiete GA (2016) Thermoelectric transport in double-Weyl semimetals. Phys Rev B 93(15):155125
55. Spivak B, Andreev A (2016) Magnetotransport phenomena related to the chiral anomaly in Weyl semimetals. Phys Rev B 93(8):085107
56. Shekhar C, Nayak AK, Sun Y et al (2015) Extremely large magnetoresistance and ultrahigh mobility in the topological Weyl semimetal candidate NbP. Nat Phys 11(8):645–649
57. Narayanan A, Watson M, Blake S et al (2015) Linear magnetoresistance caused by mobility fluctuations in n-doped Cd_3As_2. Phys Rev Lett 114(11):117201
58. Burkov A (2016) Topological semimetals. Nat Mater 15(11):1145–1148
59. Yan B, Felser C (2017) Topological materials: Weyl semimetals. Annu Rev Condens Matter Phys 8:337–354
60. Crassee I, Sankar R, Lee WL et al (2018) 3D Dirac semimetal Cd_3As_2: A review of material properties. Phys Rev Mater 2(12):120302
61. Lv BQ, Qian T, Ding H (2021) Experimental perspective on three-dimensional topological semimetals. Rev Mod Phys 93:025002
62. Zhu C, Yuan X, Xiu F et al (2017) Broadband hot-carrier dynamics in three-dimensional Dirac semimetal Cd_3As_2. Appl Phys Lett 111(9):091101
63. Weber CP, Schoop LM, Parkin SS et al (2018) Directly photoexcited Dirac and Weyl fermions in ZrSiS and NbAs. Appl Phys Lett 113(22):221906
64. Winzer T, Knorr A, Malic E (2010) Carrier multiplication in graphene. Nano Lett 10(12):4839–4843
65. Gierz I, Petersen JC, Mitrano M et al (2013) Snapshots of non-equilibrium Dirac carrier distributions in graphene. Nat Mater 12(12):1119–24
66. Gierz I, Mitrano M, Petersen JC et al (2015) Population inversion in monolayer and bilayer graphene. J Phys Condens Matter 27(16):164204
67. Sumida K, Ishida Y, Yoshikawa T et al (2019) Inverted Dirac-electron population for broadband lasing in a thermally activated p-type topological insulator. Phys Rev B 99(8):085302
68. Afanasiev A, Greshnov A, Svintsov D (2019) Relativistic suppression of Auger recombination in Weyl semimetals. Phys Rev B 99(11):115202
69. Li T, Luo L, Hupalo M et al (2012) Femtosecond population inversion and stimulated emission of dense Dirac fermions in graphene. Phys Rev Lett 108(16):167401
70. Weber CP (2021) Ultrafast investigation and control of Dirac and Weyl semimetals. J Appl Phys 129(7):070901
71. Pertsova A, Balatsky AV (2020) Dynamically induced excitonic instability in pumped Dirac materials. Ann Phys (Berl) 532(2):1900549
72. Mott NF (1961) The transition to the metallic state. Philos Mag 6(62):287–309
73. Keldysh L, Kopaev YV (1965) Possible instability of semimetallic state toward Coulomb interaction. Sov Phys Solid State 6(9):2219
74. Jérome D, Rice T, Kohn W (1967) Excitonic insulator. Phys Rev 158(2):462
75. Bucher B, Steiner P, Wachter P (1991) Excitonic insulator phase in $TmSe_{0.45}Te_0.55$. Phys Rev Lett 67(19):2717
76. Wakisaka Y, Sudayama T, Takubo K et al (2009) Excitonic insulator state in Ta_2NiSe_5 probed by photoemission spectroscopy. Phys Rev Lett 103(2):026402
77. Monney C, Battaglia C, Cercellier H et al (2011) Exciton condensation driving the periodic lattice distortion of $1T\text{-}TiSe_2$. Phys Rev Lett 106(10):106404

78. Triola C, Pertsova A, Markiewicz RS et al (2017) Excitonic gap formation in pumped Dirac materials. Phys Rev B 95(20):205410
79. Pertsova A, Balatsky AV (2018) Excitonic instability in optically pumped three-dimensional Dirac materials. Phys Rev B 97(7):075109
80. Rana F (2007) Electron-hole generation and recombination rates for Coulomb scattering in graphene. Phys Rev B 76:155431
81. Winzer T, Malić E (2012) Impact of Auger processes on carrier dynamics in graphene. Phys Rev B 85:241404
82. Tomadin A, Brida D, Cerullo G et al (2013) Nonequilibrium dynamics of photoexcited electrons in graphene: Collinear scattering, Auger processes, and the impact of screening. Phys Rev B 88:035430
83. Cayssol J, Dóra B, Simon F et al (2013) Floquet topological insulators. Phys Status Solidi RRL 7(1–2):101–108
84. Bukov M, D'Alessio L, Polkovnikov A (2015) Universal high-frequency behavior of periodically driven systems: from dynamical stabilization to Floquet engineering. Adv Phys 64(2):139–226
85. Wang Y, Claassen M, Pemmaraju CD et al (2018) Theoretical understanding of photon spectroscopies in correlated materials in and out of equilibrium. Nat Rev Mater 3(9):312–323
86. Parameswaran S, Vasseur R (2018) Many-body localization, symmetry and topology. Rep Prog Phys 81(8):082501
87. Oka T, Kitamura S (2019) Floquet engineering of quantum materials. Annu Rev Condens Matter Phys 10:387–408
88. De Giovannini U, Hübener H (2019) Floquet analysis of excitations in materials. J Phys Mater 3(1):012001
89. Rudner MS, Lindner NH (2020) Band structure engineering and non-equilibrium dynamics in Floquet topological insulators. Nat Rev Phys 2(5):229–244
90. de la Torre A, Kennes DM, Claassen M et al (2021) Colloquium: Nonthermal pathways to ultrafast control in quantum materials. Rev Mod Phys 93:041002
91. Bao C, Tang P, Sun D et al (2022) Light-induced emergent phenomena in 2D materials and topological materials. Nat Rev Phys 4:33–48
92. Oka T, Aoki H (2009) Photovoltaic hall effect in graphene. Phys Rev B 79(8):081406
93. Kitagawa T, Berg E, Rudner M et al (2010) Topological characterization of periodically driven quantum systems. Phys Rev B 82(23):235114
94. Kitagawa T, Oka T, Brataas A et al (2011) Transport properties of nonequilibrium systems under the application of light: Photoinduced quantum hall insulators without landau levels. Phys Rev B 84(23):235108
95. Inoue J, Tanaka A (2010) Photoinduced transition between conventional and topological insulators in two-dimensional electronic systems. Phys Rev Lett 105(1):017401
96. Rudner MS, Lindner NH, Berg E et al (2013) Anomalous edge states and the bulk-edge correspondence for periodically driven two-dimensional systems. Phys Rev X 3(3):031005
97. Usaj G, Perez-Piskunow PM, Torres LF et al (2014) Irradiated graphene as a tunable Floquet topological insulator[. Phys Rev B 90(11):115423
98. Perez-Piskunow PM, Usaj G, Balseiro CA et al (2014) Floquet chiral edge states in graphene. Phys Rev B 89(12):121401
99. Sentef M, Claassen M, Kemper A et al (2015) Theory of Floquet band formation and local pseudospin textures in pump-probe photoemission of graphene. Na. Commun 6(1):1–8
100. Wang YH, Steinberg H, Jarillo-Herrero P et al (2013) Observation of Floquet-Bloch states on the surface of a topological insulator. Science 342(6157):453–457
101. Mahmood F, Chan CK, Alpichshev Z et al (2016) Selective scattering between Floquet-Bloch and Volkov states in a topological insulator. Nat Phys 12(4):306–310
102. McIver JW, Schulte B, Stein FU et al (2020) Light-induced anomalous Hall effect in graphene. Nat Phys 16(1):38–41
103. Wang R, Wang B, Shen R et al (2014) Floquet Weyl semimetal induced by off-resonant light. EPL (Europhysics Letters) 105(1):17004

104. Ebihara S, Fukushima K, Oka T (2016) Chiral pumping effect induced by rotating electric fields. Phys Rev B 93(15):155107
105. Chan CK, Oh YT, Han JH et al (2016) Type-II Weyl cone transitions in driven semimetals. Phys Rev B 94:121106
106. Hübener H, Sentef MA, Giovannini UD et al (2017) Creating stable Floquet-Weyl semimetals by laser-driving of 3D Dirac materials. Nat Commun 8:13940

Chapter 2
Experimental Techniques

2.1 Angle-Resolved Photoemission Spectroscopy

For studying the equilibrium electronic structure of Dirac semimetal, angle-resolved photoemission spectroscopy (ARPES) is used which is an experimental technique for measuring the electronic structure of materials in condensed matter physics [1–5]. It is based on the photoelectric effect [6, 7] as shown in Fig. 2.1. By shining a light on the material, electrons are excited out of the material, namely, photoelectrons. By measuring the states of photoelectrons, the states of electrons inside the material can be revealed, namely, the electronic structures.

2.1.1 Basic Principles

2.1.1.1 Measurement of Electronic Dispersion

ARPES is a powerful technique for directly measuring electronic dispersion [band structure E(k)] in materials [1–5]. The principle behind it is based on the simplified three-step model [8] which is used to describe the photoemission process theoretically as shown in Fig. 2.2a. The photoemission process is divided into three steps:

The three-step model of photoemission:

1. Electrons are directly excited by photons
2. Electrons travel to the sample surface
3. Electrons go through the sample surface and into the vacuum

First, the electron is directly excited by a photon, and the energy of the electron increases by a photon energy of $h\upsilon$ while the momentum of the electron keeps the same due to the negligible momentum of photons (this is usually right for photon

© Tsinghua University Press 2023

C. Bao, *Electronic Band Structure Engineering and Ultrafast Dynamics of Dirac Semimetals*, Springer Theses, https://doi.org/10.1007/978-981-99-5325-7_2

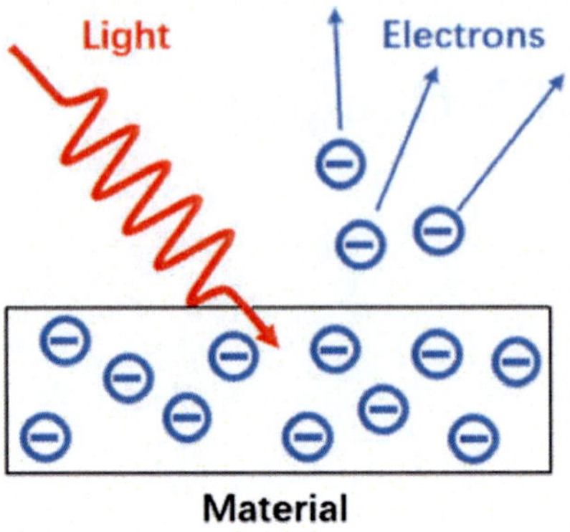

Fig. 2.1 A schematic of photoelectric effect

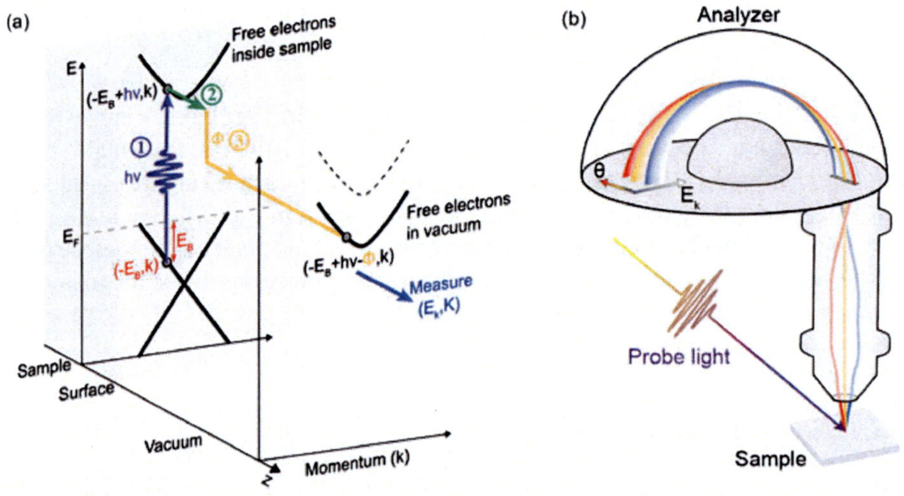

Fig. 2.2 Basic principles of ARPES. **a** Three-step model. **b** A schematic of ARPES

energy less than hundreds of eV) as indicated by the purple arrow ① in Fig. 2.2b. Secondly, the electron travels to the sample surface with the unchanged energy and momentum as indicated by the green arrow ② in Fig. 2.2a. Thirdly, the electron goes through the sample surface. Due to the energy barrier between the material and vacuum (work function Φ), the electron loses the energy of Φ. Meanwhile, because the in-plane translation symmetry is higher in vacuum than in material, the in-plane momentum is conserved, however, the out-of-plane momentum (k_z) is not conserved due to the translation symmetry breaking across the surface. Therefore, the momentum (K) and energy (E_k) of the photoelectron in vacuum are related to the momentum (k) and energy (E_B) of the electron inside the material by

$$\begin{cases} E_B = hv - \Phi - E_k \\ k_{//} = K_{//} \end{cases} \tag{2.1}$$

Therefore, by measuring the energy and momentum of photoelectrons using an electron analyzer as shown in Fig. 2.2b, we can obtain the energy and momentum of electrons inside the material, namely, the electronic dispersion $E(k)$. Due to the lack of out-of-plane momentum (k_z) in two-dimensional materials, ARPES has been widely used for studying the electronic structure of two-dimensional materials [1, 3, 5]. For three-dimensional materials, k_z becomes very important, in order to obtain k_z, another assumption is used: when the electron goes through the surface, the out-of-plane momentums are connected by the energy difference called inner potential V_0. Then, k_z can be obtained by

$$k_z = \sqrt{K_z^2 + 2mV_0/\hbar^2} \tag{2.2}$$

Considering a special case without losing generality to get more insight into the k_z measurement: for the electron at the Fermi energy and the Γ point ($k_{//} = 0$),

$$k_z = \sqrt{2m(h\upsilon - \Phi + V_0)/\hbar^2} \tag{2.3}$$

As we can see, the photon energy $h\upsilon$ is strongly linked to k_z, so the fixed photon energy corresponds to fixed k_z. In order to measure the electronic dispersion at different k_z, the photon energy needs to be changed. Such 3D measurements can be performed in the large-scale synchrotron-base ARPES station [5], which delivers widely tunable photon energy. It can be also achieved by table-top laser-based ARPES with relatively lower photon energy [9].

2.1.1.2 Measurement of Spectral Function

In addition to the simplified three-step model, a more accurate theoretical description is the direct quantum version, namely, the one-step model [10]. We assume the system remains the same when we remove one electron (the sudden approximation), the photoemission process corresponds to the direct electronic transition from the initial state to the final state and the intensity can be expressed by

$$I(\mathbf{k}, E) \propto f(E)|M_{ij}|^2 A(\mathbf{k}, E) \tag{2.4}$$

where $f(E)$ is the Fermi-Dirac function, M_{ij} is the matrix element of the transition, and $A(\mathbf{k}, E)$ is the single-particle spectral function that describes the transition possibility of removing one electron from the system. The spectral function can be expressed instead of self-energy Σ,

$$A(\boldsymbol{k}, E) = \frac{1}{\pi} \frac{\mathrm{Im}\Sigma}{(E - E^0(\boldsymbol{k}) - \mathrm{Re}\Sigma)^2 + (\mathrm{Im}\Sigma)^2} \tag{2.5}$$

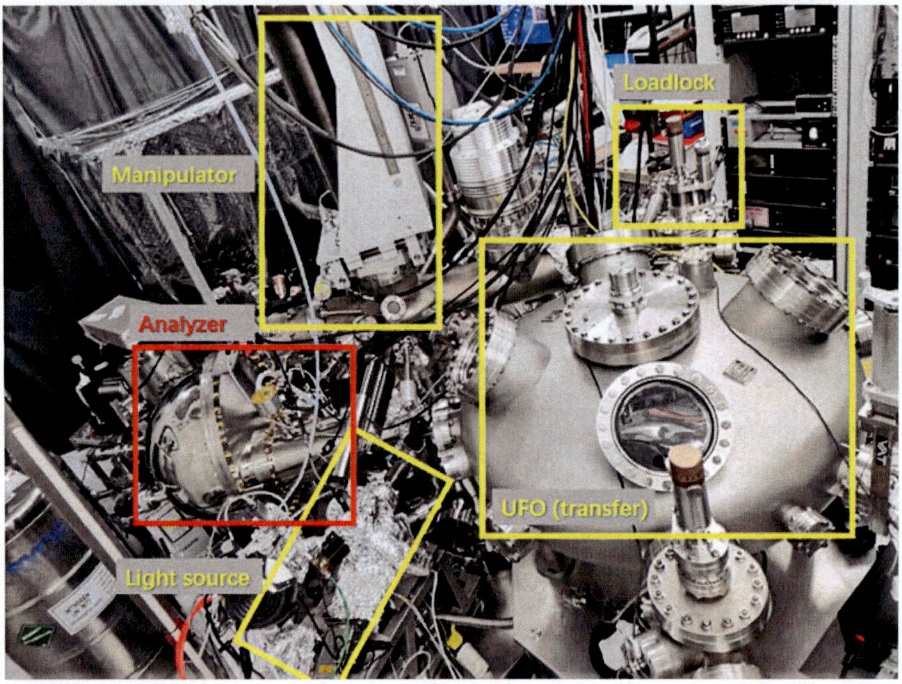

Fig. 2.3 The photograph of a typical ARPES system

where the real and imaginary parts of the electron self-energy correspond to the band dispersion and width respectively. Therefore, the electronic dispersion $E(k)$ and lifetime are be measured by ARPES.

More importantly, the self-energy not only reflects the basic electronic structure but also the many-interaction information, for example, electron-phonon interaction and electron-electron interaction. In an interacting electron system, the self-energy $\mathrm{Re}\Sigma$ derives from the intrinsic band dispersion (bare band) and $\mathrm{Im}\Sigma$ also changes due to the scattering.

2.1.2 Instruments

Experimentally, in order to measure the energy and momentum of photoelectrons, the ARPES system is consist of a hemisphere analyzer, a manipulator, light source, vacuum chamber, and other supporting systems as shown in Fig. 2.3. In the following, the instruments will be introduced in detail.

The most important component is the hemisphere analyzer, which can be divided into two parts: electronic lens and hemisphere chamber as shown in Fig. 2.4. First, the photoelectrons fly into the electronic lens which focuses the electrons emitted from

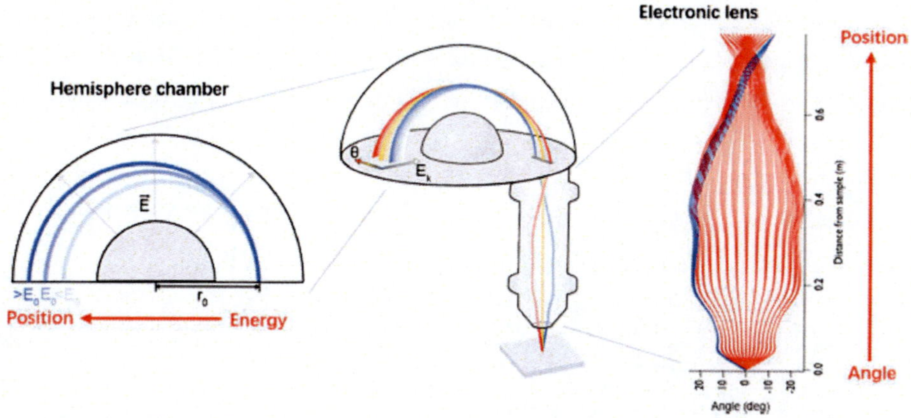

Fig. 2.4 Schematics for hemisphere analyzer and electron trajectories inside the hemisphere chamber and electronic lens

the same angle into the same position, so the angle information is converted into the position information and can be detected directly by the CCD camera. After that, the focused electrons fly into the hemisphere chamber which is used to resolve the energy. As we can see in Fig. 2.4, only the electron with a specific energy $E_0 = r_0 \boldsymbol{E}e/2$ can move along a circle exactly. The electrons with energy larger or smaller than E_0 shift outside or inside respectively, so the energy information is also converted into position information and can be directly measured. Therefore, by measuring the electronic signal in a two-dimensional plane through a CCD camera, both energy and angle information can be obtained.

However, the measured angle is only one-dimensional (θ_x) as indicated by the red arrow in Fig. 2.5a, in order to access the full two-dimensional angle space, we need to rotate the sample by θ_y as indicated by the blue arrow. Therefore, the sample is mounted in a motorized manipulator which allows not only rotation in three directions for measuring the full angle space, but also translation along with three directions for measuring different positions on the sample and adjusting the focus distance as shown in Fig. 2.5b. In addition, the manipulator also provides a well-controlled sample temperature, so that the ARPES measurements can be performed from several Kelvins to room temperature, which is very important for studying the materials with phase transition.

Therefore, by using the combination of the hemisphere analyzer and manipulator, the energy and angle of photoelectrons can be measured directly. Then, we can convert it to momentum for the electronic structure E(k) by

$$\begin{cases} K_x = \sqrt{\frac{2mE_k}{\hbar^2}} \cos\theta_y \sin\theta_x \\ K_y = \sqrt{\frac{2mE_k}{\hbar^2}} \sin\theta_x \\ K_z = \sqrt{\frac{2mE_k}{\hbar^2}} \cos\theta_y \cos\theta_x \end{cases} \tag{2.6}$$

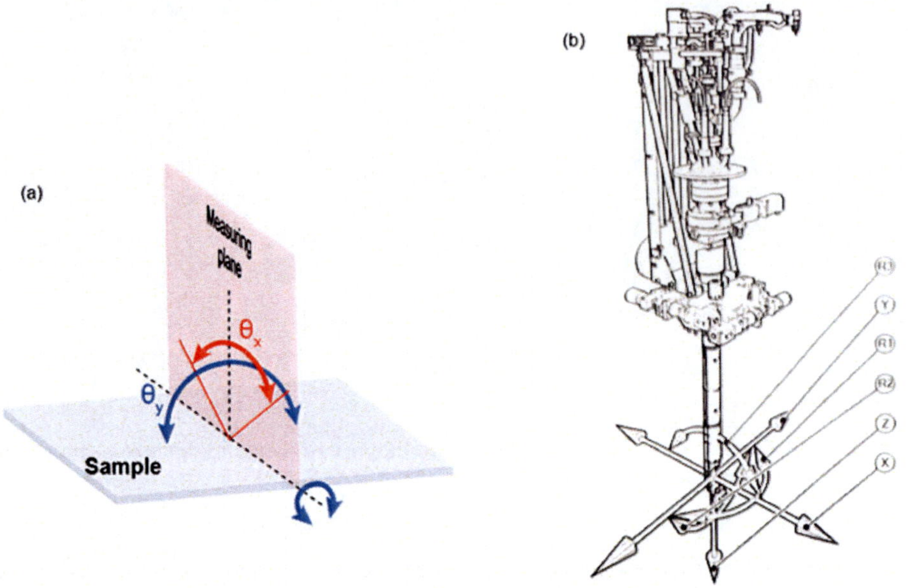

Fig. 2.5 Schematics for sample rotation (**a**) and the manipulator (**b**)

In addition to the hemisphere analyzer and manipulator, the light source is also important, which is used to generate the photoelectrons. Because the work function of materials is typically around 4–5 eV, a light source with photon energy larger than the work function is needed. The typically used light source includes a gas-discharge lamp represented by the Helium lamp, large-scale synchrotron radiation, and table-top coherent ultraviolet (UV) or vacuum ultraviolet (VUV) laser. In addition, other supporting systems including loadlock, transfer chamber, and preparation chamber are used for loading the sample from the atmosphere into the ultrahigh vacuum, transferring the sample between different chambers, and cleaning the sample surface as shown in Fig. 2.3.

2.2 Time- and Angle-Resolved Photoemission Spectroscopy

ARPES shows powerful capabilities for measuring the equilibrium electronic structure, by combing the pump-probe technique and ARPES, time- and angle-resolved photoemission spectroscopy (TrARPES) can measure the non-equilibrium electronic structure in the time domain, which is a powerful technique to study the light-matter interaction in solid [5, 11], including the non-equilibrium electronic structure of Dirac semimetal.

2.2.1 Basic Principles

TrARPES is a powerful technique to resolve unoccupied states. From the view of electronic structure, the pump light excites the electrons from occupied states below Fermi energy to unoccupied states above Fermi energy as shown in Fig. 2.6. Then, the probe light can detect not only the electrons in occupied states but also in unoccupied states. To obtain enough strong signals for the unoccupied states, strong pump light is needed and an ultrashort pulsed laser (from tens to hundreds of fs) is typically used. However, after the transient photoexcitation, the electrons in unoccupied states are not stable and will relax back eventually, so the pump and probe pulse should coincide in time in order to see the unoccupied states. For example, by using TrARPES, the topological surface states above Fermi level in topological insulator Sb_2Te_3 can be clearly observed as shown in Fig. 2.7.

TrARPES is also a powerful technique to measure the ultrafast dynamics of photocarriers. By continuously varying the delay time between pump and probe pulse, the relaxation process can be directly visualized as shown in Fig. 2.8. Therefore, how the electrons are excited and relax between different electronic bands and the lifetimes of photocarriers in different bands can be revealed, which provides insight for understanding the relaxation mechanism and important parameters for high-speed photoelectric device applications.

In addition to probing the unoccupied states and ultrafast dynamics, an even more fascinating capability of TrARPES is probing the light-induced band engineering on an ultrafast time scale in a nonthermal way, thus providing opportunities for exploring the new phase of mater which has no counterpart in equilibrium states. For example, band gap opening and topological transition through Floquet engineering in 2D Dirac system [12–17].

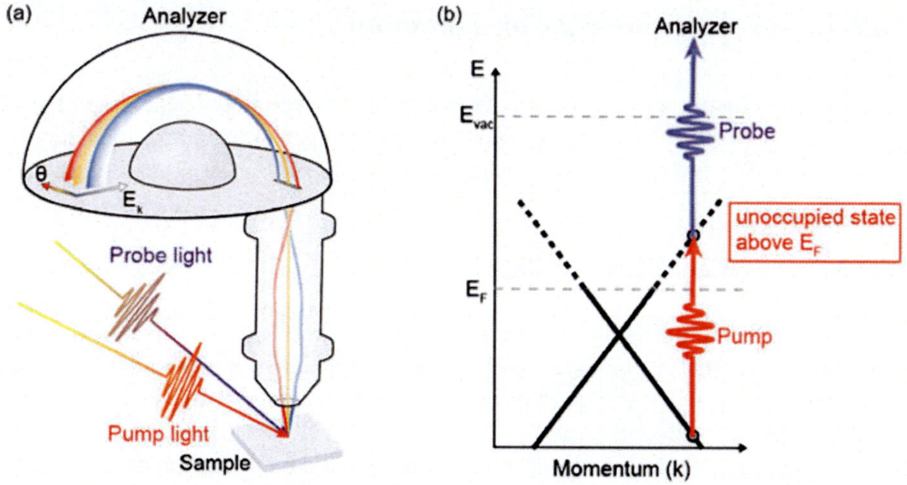

Fig. 2.6 Schematics for TrARPES setup and principles. **a** Schematics for TrARPES. **b** Band diagram for pump and probe process in TrARPES for measuring the unoccupied states

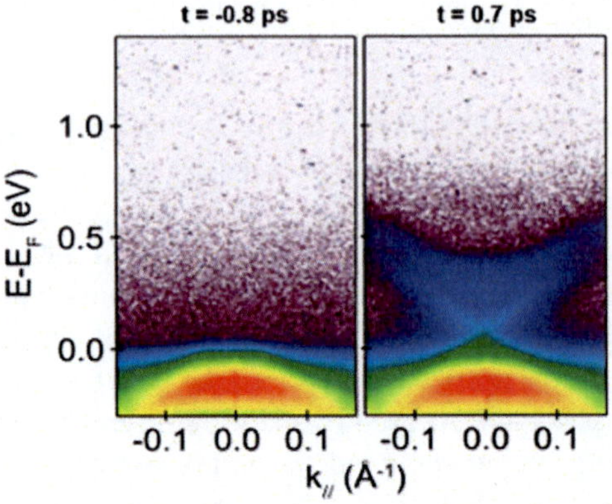

Fig. 2.7 Revealing unoccupied states of topological insulator Sb_2Te_3 by TrARPES

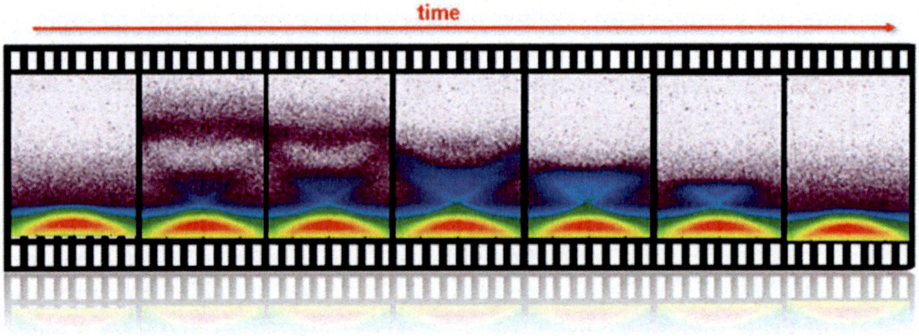

Fig. 2.8 Revealing temporal evolution of unoccupied states of topological insulator Sb_2Te_3 by TrARPES

2.2.2 *Special Technical Considerations*

The detailed instruments of TrARPES will be introduced in the next chapter as one of the main works of this thesis. Here, only some special technical considerations for TrARPES, which are not serious in conventional ARPES, will be discussed.

2.2.2.1 Space and Time Overlap

In TrARPES, the pump and probe beams should overlap perfectly in both space and time. The time is one dimension and the space is two dimensions, it is almost impossible to look for the overlap in all three dimensions at the same time and a special strategy is needed.

The strategy is to ensure the space overlap first and then scan the time dimension to look for the TrARPES signal. We use a copper plate covered with the phosphorus attached to the manipulator to see both pump and probe beams. The beam spots

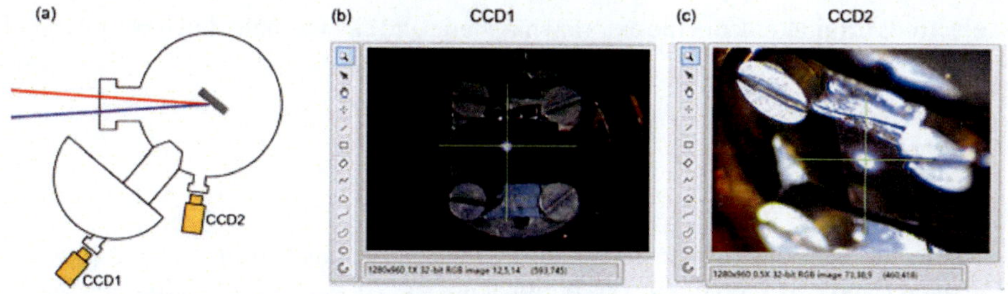

Fig. 2.9 Looking for space overlap. **a** Schematics of the CCD camera system for space overlap. **b**, **c** Images from two CCD cameras to show the beam spot on the sample

are monitored by two CCD cameras as shown in Fig. 2.9. Two beams are adjusted to overlap with each other directly by the alignment mirrors. The beam position is marked in both CCD cameras. Then, we move the sample into the position and adjust the sample position very carefully to ensure it coincides with the markers in both CCD cameras as shown in Fig. 2.9b, c. For the sample on which you can find the beam spot (for example, large-size thin film), we adjust the sample position to ensure the beam spot on the sample coincides with the markers. For the sample on which we can not see the beam spot, we look for some sharp features to coincide with the markers. Then, the pump and probe beams overlap on the sample spatially.

Then we adjust the time overlap. We directly measure the optical path of the pump and probe beams by a ruler, which gives an estimation of the time zero point with an uncertainty of less than 1 cm. Then, we scan the delay time nearby to look for the TrARPES signal. Actually, in the daily operation, it is not necessary to check the time zero point which is quite stable.

2.2.2.2 Low-Energy Cutoff and Work Function

In TrARPES measurements, the photon energy is typically very low (around 6 eV). When we measure the photoelectrons using very low-energy photons, the low-energy cutoff is another common issue. Such a low-energy cutoff limits the measurement energy range. It shows different values for different materials as summarized in [18], because it is directly linked to the work function of materials. For photon energy of $h\nu$, the lowest electron that can be excited out of the sample is $h\nu - \Phi$, which set the low-energy cutoff.

2.2.2.3 Electric Field Distortion

Working with low photon energy, the photoelectrons also have very low kinetic energy which needs special considerations for the distortion of photoelectrons compared to the conventional ARPES with high photon energy. The distortion of photoelectrons originates from the external magnetic and electric field, the induced impulse can be expressed:

$$\begin{cases} I_B = Bev \cdot \frac{l}{v} = Bel \\ I_E = Ee \cdot \frac{l}{v} = Eel/v, \end{cases} \tag{2.7}$$

where the e is the charge of electrons, l is the distance from the sample to the analyzer and v is the velocity of photoelectrons. As we can see, when the velocity of electrons decreases, the magnetic field induced impulse remains the same, while the electric field induced impulse increase rapidly. Therefore, for the low-energy electrons, the electric field dominates the distortions and the external electric field should be carefully avoided, for example, do not use any wobble stick inside the measuring chamber.

However, there is an unavoidable origin of the electric field is from the sample itself. Because different materials possess different work functions, when they contact each other, a potential difference appears to balance their different Fermi levels, so-called contact potential. In TrARPES measurement, the sample is different from the sample holder, so there exists contact potential between them and strongly affects the trajectory of photoelectrons. Moreover, the position is closer to the sample edge, the distortion becomes stronger [18]. Therefore, a large-size sample and measuring at the center of the sample are very important to reduce the distortion.

References

1. Damascelli A, Hussain Z, Shen ZX (2003) Angle-resolved photoemission studies of the cuprate superconductors. Rev Mod Phys 75(2):473–541
2. Zhou X, He S, Liu G et al (2018) New developments in laser-based photoemission spectroscopy and its scientific applications: a key issues review. Rep Prog Phys 81(6):062101
3. Lv B, Qian T, Ding H (2019) Angle-resolved photoemission spectroscopy and its application to topological materials. Nat Rev Phys 1(10):609–626
4. Lv BQ, Qian T, Ding H (2021) Experimental perspective on three-dimensional topological semimetals. Rev Mod Phys 93:025002
5. Sobota JA, He Y, Shen ZX (2021) Angle-resolved photoemission studies of quantum materials. Rev Mod Phys 93:025006
6. Hertz H (1887) Ueber einen einfluss des ultravioletten lichtes auf die electrische entladung. Ann Phys (Berl) 267(8):983–1000
7. Einstein A (1905) Über einem die erzeugung und verwandlung des lichtes betreffenden heuristischen gesichtspunkt. Ann Phys (Berl) 4
8. Hüfner S (2013) Photoelectron spectroscopy: principles and applications. Springer Science & Business Media
9. Jiang R, Mou D, Wu Y et al (2014) Tunable vacuum ultraviolet laser based spectrometer for angle resolved photoemission spectroscopy. Rev Sci Instrum 85(3):033902
10. Mitchell K (1934) The theory of the surface photoelectric effect in metals I. Proc R Soc London Ser 146(857):442–464
11. Smallwood CL, Kaindl RA, Lanzara A (2016) Ultrafast angle-resolved photoemission spectroscopy of quantum materials. Europhys Lett 115(2):27001

12. Wang YH, Steinberg H, Jarillo-Herrero P et al (2013) Observation of Floquet-Bloch states on the surface of a topological insulator. Science 342(6157):453–457
13. Mahmood F, Chan CK, Alpichshev Z et al (2016) Selective scattering between Floquet-Bloch and Volkov states in a topological insulator. Nat Phys 12(4):306–310
14. Oka T, Kitamura S (2019) Floquet engineering of quantum materials. Annu Rev Condens Matter Phys 10:387–408
15. De Giovannini U, Hübener H (2019) Floquet analysis of excitations in materials. J Phys Mater 3(1):012001
16. Rudner MS, Lindner NH (2020) Band structure engineering and non-equilibrium dynamics in Floquet topological insulators. Nat Rev Phys 2(5):229–244
17. Bao C, Tang P, Sun D et al (2022) Light-induced emergent phenomena in 2D materials and topological materials. Nat Rev Phys 4:33–48
18. Fero A, Smallwood C, Affeldt G et al (2014) Impact of work function induced electric fields on laser-based angle-resolved photoemission spectroscopy. J Electron Spectr Relat Phenom 195:237–243

Chapter 3
Development of Novel TrARPES with Tunable Probe Photon Energy for 3D Quantum Materials

3.1 Research Motivation

Revealing the ultrafast dynamics of quantum materials can provide critical information into the non-equilibrium state and possible light-induced emerging phenomena [1]. Time- and angle-resolved photoemission spectroscopy (TrARPES) [2, 3] has been a powerful technique for capturing the ultrafast carrier dynamics and photo-induced phase transitions with energy-, momentum- and time-resolution. In the past decades, major progress has been made in TrARPES instrumentation, including extending the pump wavelength to mid-infrared [4] or terahertz [5], improving the energy resolution or time resolution [6–9], etc. Developments in the probe source using high harmonic generation (HHG) [10–12], or free electron laser [13] with a higher probe photon energy has also been achieved. While the power of TrARPES has been demonstrated in various quasi-two-dimensional materials such as high-temperature superconductors, topological insulators and graphene [1–3] to extract the ultrafast dynamics, so far, TrARPES studies of 3D materials such as 3D Dirac or Weyl semimetals, where intriguing light-induced dynamics and phase transitions have been predicted [14–17], have been restricted. The challenge for TrARPES measurements of these 3D materials [18] is that the Dirac or Weyl nodes exist only in isolated out-of-plane momentum (k_z) points, and highly tunable photon energy is required to access the dispersions of interests at different k_z with high precision.

The probe beam in TrARPES systems is typically generated by fourth harmonic generation (FHG) using nonlinear optical crystals β-BaB_2O_4 (BBO) with a cut-off photon energy of 6.3 eV [6]. Nonlinear optical crystal $KBe_2BO_3F_2$ (KBBF) crystal with prism-coupled technique (KBBF-PCT) can be used to generate high-brightness table-top vacuum ultraviolet (VUV) light source [19–24] up to 7.0 eV with excellent performance for ARPES measurements [25–29]. A continuously tunable probe photon energy of 5.9–7.0 eV has been demonstrated in static ARPES measurements [27]. KBBF-based TrARPES measurements have been performed at a fixed photon

© Tsinghua University Press 2023

C. Bao, *Electronic Band Structure Engineering and Ultrafast Dynamics of Dirac Semimetals*, Springer Theses, https://doi.org/10.1007/978-981-99-5325-7_3

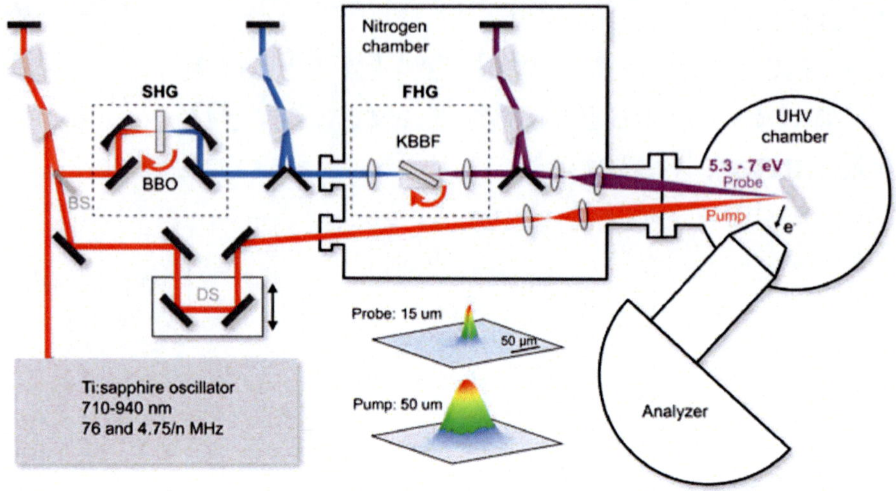

Fig. 3.1 A schematic layout of TrARPES system with tunable vacuum ultraviolet laser. BS: beam splitter, DS: delay stage, UHV: ultra-high vacuum. Transverse intensity profiles of pump and probe beams. The beam sizes are defined by full width at half maximum (FWHM). Reproduced, with permission from [31]

energy with a time resolution of ~1 ps [30]. Implementing TrARPES with a highly tunable probe wavelength while achieving a high time resolution is critical for further extending the ultrafast dynamics studies to 3D materials.

3.2 TrARPES with Tunable Probe Photon Energy

3.2.1 Experimental Setup

Figures 3.1 and 3.2 shows an overview of the TrARPES system. The fundamental beam (FB) with a wavelength of 710–940 nm is generated by a Ti: sapphire oscillator (Coherent Mira HP) with an output power of 2–4 W and a repetition rate of 76 MHz. The repetition rate can be reduced to 4.75/n (n is an integer) MHz by an external pulse picker. Such flexibility allows choosing an optimum repetition rate with high data acquisition efficiency while avoiding sample damage problems induced by average heating. The laser beam is firstly compressed by a prism pair compressor to compensate for the chirp from the laser cavity. A beam splitter splits the FB into two branches: 50% for the pump beam, and the other 50% is used to generate the probe beam through FHG. Compared to an amplifier which gives larger pulse energy, here a higher fraction of the FB is needed to generate the high-brightness probe beam for high measurement efficiency.

Fig. 3.2 Photographs of the TrARPES system with tunable probe photon energy

The pump beam goes through a delay stage which is used to change the delay time between the pump and probe pulses. Several silver mirrors are used to steer the beam onto the sample in the ultra-high vacuum (UHV) chamber for photoexcitation. Before entering the UHV chamber, the pump beam is focused by a lens pair with a focal length of f = 50 and 200 mm to obtain a beam size of 50 µm as shown in Fig. 3.1. A maximum pump fluence of 1 mJ/cm^2 can be applied on the sample surface.

The other 50% of the FB is used to generate the probe beam. It is first focused onto a second harmonic generation (SHG) BBO crystal with type I phase matching ($\theta = 29.2°$, thickness 0.5 mm) by an off-axis parabolic mirror (aluminum coating, f = 50 mm). The choice of an off-axis parabolic mirror instead of a lens is to reduce the possible chirp. The output laser beam is then collimated by another off-axis parabolic mirror and sent to a prism pair compressor to achieve optimum pulse duration for the second harmonic (SH) beam at the KBBF crystal. We would like to point out that compared to the BBO-base FHG where the FHG is achieved by passing the SH through the BBO crystal directly without generating significant chirp, compression of SH is not so critical, while in the current setup, the SH passes through a large-size prism made of fused silica with a length of 15 mm before the KBBF crystal, and the chirp is significant. For example, a 50 fs SH pulse at 400 nm is elongated to 95 fs after passing through the prism. Therefore, pre-compensation for the chirp is required to achieve the optimal pulse duration for the fourth harmonic (FH) laser. The fundamental and second harmonics lasers are separated by the first prism in the compressor, where the residual fundamental laser beam is dumped.

The SH laser beam is focused by a lens (ultra-violet fused silica, f = 50 mm) onto the KBBF-PCT device [22] with type I phase matching (thickness of KBBF is 1 mm) to generate the probe beam as shown in Fig. 3.2b. The SH and FH laser beams are separated by the prism behind the KBBF crystal, to dump the residual SH laser beam. The FH laser beam is subsequently collimated by a CaF_2 lens with f = 100 mm to ensure the long-distance propagation. Then it goes into a prism pair compressor for pulse duration compression and is finally focused by a lens pair (CaF_2, f = 50 and 150 mm) to obtain a probe beam size of 15 um on the sample surface. All the FH-related components are placed inside a nitrogen-purged chamber to avoid absorption of VUV laser for photon energy above 6.7 eV by the oxygen [32], water vapor, carbon oxide, and organic molecular in the air. The photon energies of the pump and probe beams are related by $h\nu_{probe} = 4h\nu_{pump}$ and cannot be tuned independently. However, the change in the pump photon energy is much smaller (from 1.33 to 1.75 eV), and the effect on the dynamics is negligible unless there is any resonant pumping.

The sample is mounted on a motorized manipulator in an ultrahigh vacuum (UHV) chamber with a vacuum better than 5×10^{-11} Torr. The manipulator has six degrees of freedom, allowing full control of the sample position and orientation. The sample is electrically insulated from the ground and connected to a picoammeter (Keithley 6485) to monitor the photocurrent. The energy and momentum of photoelectrons are measured by an electron analyzer (Scienta, DA30-L-8000).

3.2.2 *Tunable Probe Photon Energy*

The tunable probe photon energy is realized by adjusting the wavelength of the FB and corresponding phase-matching angle of nonlinear BBO and KBBF crystals. To make sure the phase matching in the entire wavelength range, the BBO is installed on a large-range rotation stage to allow rotation as indicated by the red circular arrow in Fig. 3.3a, and the KBBF-PCT device is installed on a piezoelectric driven large-range rotation stage as indicated by the red arrow in Fig. 3.3b.

3.2.2.1 High-Brightness Probe Beam Covering 5.3–7.0 eV

The input FB beam before the SHG crystal is 160–800 mW at a wavelength of 710–940 nm as shown in Fig. 3.3c and the output power of the SH laser is 9–250 mW at a wavelength of 355–470 nm as shown in Fig. 3.3d. The corresponding SH efficiency is 6–30%. The power of final FH laser is 3–430 μW, which corresponds to a photon flux of 4×10^{12}–4×10^{14} s^{-1} at photon energy of 5.3–7.0 eV (235–177.5 nm) as shown in Fig. 3.3e, which is much higher than typical HHG light source ($<10^{11}$ s^{-1}) [33]. Therefore, a high-brightness probe laser beam with a tunable photon energy of 5.3–7.0 eV is achieved for TrARPES measurements. We note that for typical BBO-based TrARPES, the highest photon energy is 6.05 eV in the $2\omega + 2\omega$ scheme or 6.3 eV in the $\omega + 3\omega$ scheme [6]. By using KBBF-PCT, we can extend it to cover the previously inaccessible probe photon energy range of 6.3–7.0 eV. In addition to accessing a larger range of k_z momentum, the higher photon energy also allows probing a larger energy range below the Fermi energy.

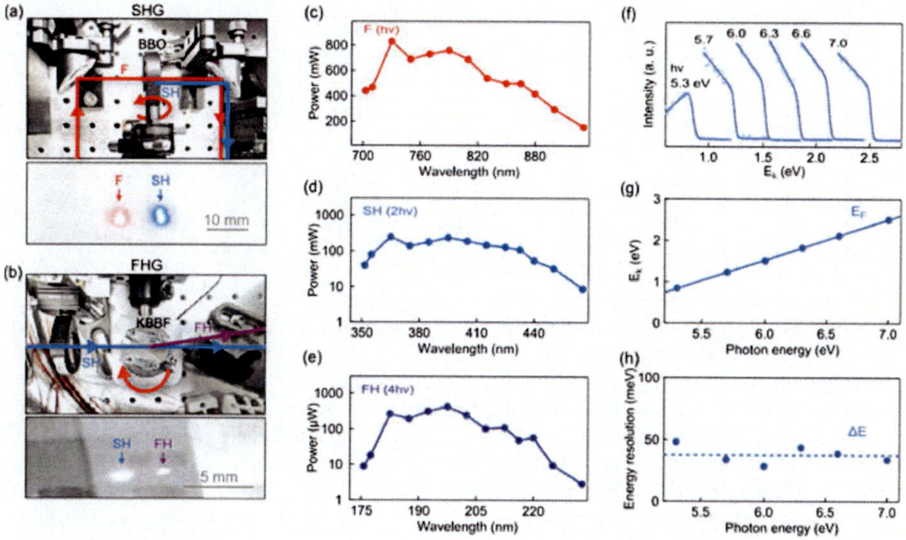

Fig. 3.3 Probe photon flux and energy resolution. **a** Photos of SHG setup and beam spots on white paper. **b** Photos of FHG setup and beam spots on the glass. **c** Fundamental laser power for SHG input as a function of fundamental wavelength. **d** Generated SH laser power as a function of SH wavelength. **e** Generated FH laser power as a function of FH wavelength. **f** Integrated EDCs on polycrystalline copper using different photon energies at 80 K. **g** Extracted kinetic energy of Fermi energy as a function of photon energy with a linear fitting. **h** Extracted energy resolution at different photon energies. The dashed line indicates the average level. Reproduced, with permission from [31]

3.2.2.2 Energy Resolutions

The energy resolution at different probe photon energies is measured on a polycrystalline copper at 80 K. The angle-integrated energy distribution curves (EDCs) for photon energies from 5.3 to 7.0 eV are shown in Fig. 3.3f. The Fermi energy (E_F) and energy resolution (ΔE) can be extracted by fitting EDCs with Fermi-Dirac function.

$$f(E) = 1/(1 + \mathrm{e}^{\frac{E-E_F}{\Delta_{wid}/4}}), \tag{3.1}$$

where Δ_{wid} corresponds to the width of the Fermi edge and is related to ΔE by $\Delta_{wid} = \sqrt{\Delta_E^2 + \Delta_T^2}$. Here $\Delta_T = 4k_B T = 28$ meV at the measurement temperature of T = 80 K is induced by the thermal broadening (defined by 12–88% of the edge), which could be reduced when cooling down to a lower temperature, and ΔE is the energy resolution with contributions from the probe source, analyzer, and sample broadening. The extracted Δ_{wid} from the fitting is 40–56 meV, and the energy resolution after subtracting the contribution from the thermal broadening Δ_T is 29–48 meV for the entire photon energy range as shown in Fig. 3.3h. The energy resolution is limited by the time resolution through the uncertainty principle and could be

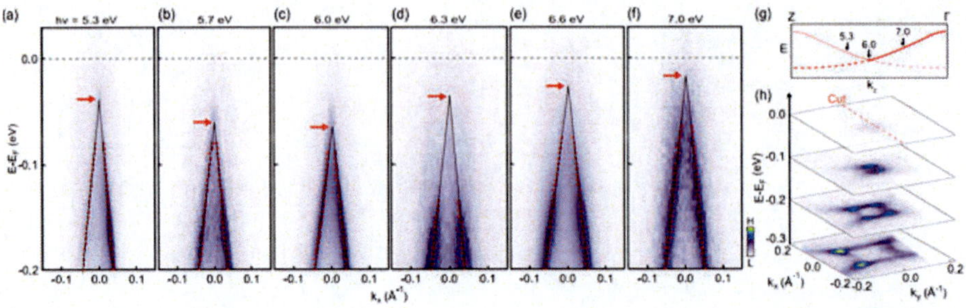

Fig. 3.4 The demonstration of the tunable photon energy on the $ZrTe_5$ single crystal. **a–f** Dispersion images along the direction as indicated in **h** using different photon energies. The red dots are extracted from momentum distribution curves peaks, and the black lines are the extrapolated dispersion. The red arrows indicate the top of the valence band. **g** The schematic band dispersion along k_z direction. The red curve is original dispersion and the light red curve is folded dispersion. The black arrows indicate the corresponding k_z for photon energies of 5.3, 6.0 and 7.0 eV. **h** Intensity maps at different energies using a photon energy of 7.0 eV. The measurement temperature is 80 K. Reproduced, with permission from [31]

improved by sacrificing the time resolution, for example, a higher energy resolution of 16 meV [34] can be achieved by compromising the time resolution.

3.2.2.3 Demonstrations on 3D Topological Material $ZrTe_5$

The suitability of such a highly tunable probe photon energy source for 3D quantum materials is demonstrated on $ZrTe_5$, which is a topological material [35, 36] with strong k_z dispersion [37–40]. Figure 3.4a–f shows dispersion images measured along the Γ-X direction (indicated by the dashed line in Fig. 3.4h) at a few selected photon energies from 5.3 to 7.0 eV. This photon energy range corresponds to $k_z = 0.70 - 0.87$ c^* [37], covering 34% of Γ - Z (half of the k_z Brillouin zone) as shown in Fig. 3.4g. For typical 3D topological materials, the lattice constant is 7.25 Å for $ZrTe_5$ [37], 9.7 Å for Na_3Bi [41], 12.6 Å for Cd_3As_2 [42] and 11.6 Å for TaAs [43], so the tunable photon energy of 5.3–7.0 eV can cover around 34%-60% of the half of the k_z Brillouin zone for typical 3D topological materials (assuming an inner potential of 7.5 eV [37]).

A significant change is clearly identified in the top of the valence band as indicated by red arrows in Fig. 3.4a–f, showing strong k_z dispersion. The shift of the top of the valence band toward the Fermi energy from photon energy of 6.0–7.0 eV is consistent with previous reports [40] due to the strong k_z dispersion along the Γ—Z direction as illustrated schematically in Fig. 3.4g. From 6.0 to 5.3 eV, the top of the valence band moves up, possibly due to the 1/2 reconstruction induced band folding along the k_z direction [37] as indicated by the light red curve in Fig. 3.4g.

3.2.3 *Considerations for the Time Resolution*

In order to capture the ultrafast electronic dynamics, a high time resolution is required. For example, typical electronic excitation or relaxation through electron-electron interaction occurs in several to tens of femtoseconds, and electron-phonon interaction occurs in hundreds of femtoseconds to several picoseconds [44, 45]. In addition, light-induced transient phenomena also emerge in a short temporal window within hundreds of femtoseconds, such as electronic oscillation induced by coherent phonon [45] and Floquet band engineering [4]. Therefore, achieving a high time resolution in TrARPES is critical for revealing electronic dynamics and light-induced exotic quantum phenomena. In the following section, the key factors for achieving high time resolution are discussed in detail.

3.2.3.1 Fourier Transform Limit

The time resolution of TrARPES is determined by the cross-correlation between pump and probe laser pulse:

$$\Delta t_{TrARPES} = \sqrt{(\Delta t_{pump})^2 + (\Delta t_{probe})^2}. \tag{3.2}$$

Therefore, the pulse durations of both pump and probe laser beams need to be optimized to reach a high time resolution.

The pulse duration is intrinsically connected to its bandwidth according to the uncertainty principle (for Gaussian pulse) [46]:

$$\Delta t \times \Delta \nu > \frac{2 \ln 2}{\pi}. \tag{3.3}$$

A wide enough bandwidth $\Delta \nu$ is therefore required for short laser pulses, and this sets a lower limit for the pulse duration which is Fourier-transform-limited pulse duration. For example, a pulse duration of 50 fs requires a bandwidth exceeding 20.8 nm at 820 nm.

3.2.3.2 Chirp Induced by Optical Components

The pulse duration can be elongated due to the chirp induced by the optical components, e.g. lens, crystals and windows, etc. Chirp refers to the separation of different wavelengths in the time domain after propagating through an optical component. It originates from group velocity dispersion (GVD) [47],

$$GVD = \partial(1/v_g)/\partial\omega = \partial^2 k/\partial\omega^2. \tag{3.4}$$

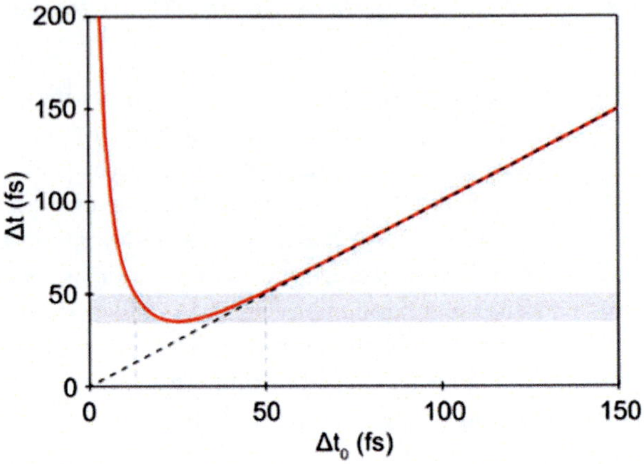

Fig. 3.5 Pulse duration of pump laser beam as a function of input laser pulse duration Δt_0 after adding a chirp with GDD = 224 fs^2. Reproduced, with permission from [48]

When the pulse travels by a distance of l, the accumulated group delay dispersion (GDD) is

$$GDD = l \times GVD \tag{3.5}$$

and this leads to broadening of the pulse from Δt_0 (without chirp) to

$$\Delta t = \Delta t_0 \sqrt{1 + \left(\frac{4 \ln 2 \times GDD}{\Delta t_0^2}\right)^2}. \tag{3.6}$$

For example, for the pump, a beam splitter (UVFS, thickness 3 mm), a lens (UVFS, thickness 2.4 mm) before the ARPES chamber, and a window (LiF, thickness 2.5 mm) on the ARPES chamber are the main sources of chirp, which contribute an overall GDD of 224 fs^2. Figure 3.5 shows the calculated pulse duration of the pump beam after these optical components as a function of input pulse durations Δt_0. The calculation shows that the shortest input pulse does not necessarily lead to the best time resolution. This can be explained as follows: the shorter pulse requires a broader bandwidth which leads to a stronger chirp.

The chirp can be compensated by a prism pair (or negative GDD mirrors or gratings) which gives a negative GDD [49]:

$$GDD_{prism} = \frac{4\lambda^3}{\pi c^2}\left(-\left(\frac{dn}{d\lambda}\right)^2 L + \frac{d^2 n}{d\lambda^2} D\right), \tag{3.7}$$

where L is the separation between the prisms. Two prism pairs are implemented to compensate for the chirp in the FB and FH (see Fig. 3.1).

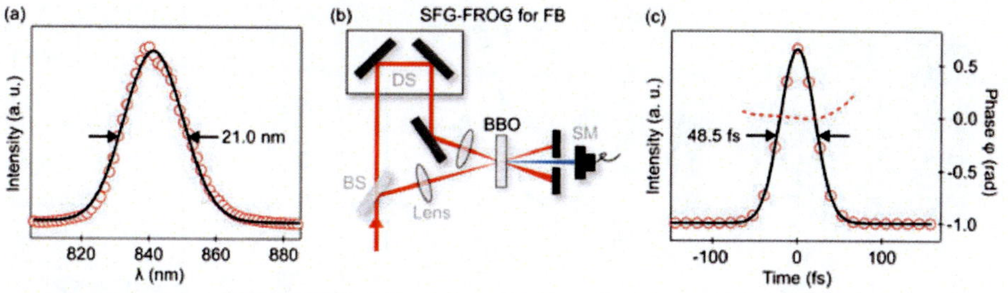

Fig. 3.6 Characterization and optimization of the FB. **a** Spectrum of the FB. **b** A schematic of the diagnosis system using SFG. The SFG signal is measured by a spectrometer. **c** Intensity and phase of FB retrieved from FROG measurement in the time domain, which gives a pulse duration of 48.5 ± 0.1 fs for the FB. Reproduced, with permission from [48]

3.2.3.3 Pulse Duration Diagnostics

In order to evaluate the effect of the compression, characterization schemes using sum-frequency generation (SFG) and difference frequency generation (DFG) have also been added to characterize the pulse duration of the FB and SH.

Here a frequency-resolved optical gating (FROG) [47] using sum-frequency generation (SFG, ω+ω in this case) is used for the diagnostics of FB (Fig. 3.6b). The pulse intensity and phase can be retrieved by iteratively searching the signal solution $E(t, \tau)$ satisfying two constraints: $E(t, \tau) = E(t)E(t - \tau)$; $I_{FROG} = |\int_{\infty}^{\infty} E(t, \tau)e^{-i\omega t}dt|^2$ [50]. After optimization of the compressor, the retrieved pulse in time domain shows a pulse duration of $\Delta t_F = 48.5 \pm 0.1$ fs in Fig. 3.6c. In addition, the retrieved phase (dashed curve in Fig. 3.6c) shows negligible dependence on time suggesting negligible chirp [47]. Therefore, the spectrum together with the SFG-FROG confirms that the FB satisfies the bandwidth, pulse duration and no chirp requirements.

Here we use a home-built FROG with a difference frequency generation (DFG) scheme to characterize SH as shown in Fig. 3.7a, b. The SH and FB are picked up from the main optical path by two flip mirrors and sent through an SHG BBO crystal (0.1 mm thick, thinner detection BBO than the generation BBO is required to obtain the intrinsic pulse duration) to generate a DFG signal at a frequency of 2ω-ω, which is measured by a power meter. By adjusting the delay time between two beam paths, temporal information is obtained. The cross-correlation between SH and FB shows a correlation width of 66 ± 2 fs in Fig. 3.7d, implying an SH pulse duration of 46 ± 3 fs. This is also in agreement with SH bandwidth of 7.3 ± 0.1 nm (Fig. 3.7a), which corresponds a Fourier-transform-limited pulse duration of 36 fs. To further confirm that there is no chirp for the SH, the FROG-DFG image showing the wavelength-dependent DFG signal is measured by a spectrometer and shown in Fig. 3.7e. We note that a chirp pulse means a change of wavelength in the propagation of the pulse as schematically shown in Fig. 3.7g. Since the fundamental beam can be treated as a perfect reference without any chirp, the chirp of the SH can be directly

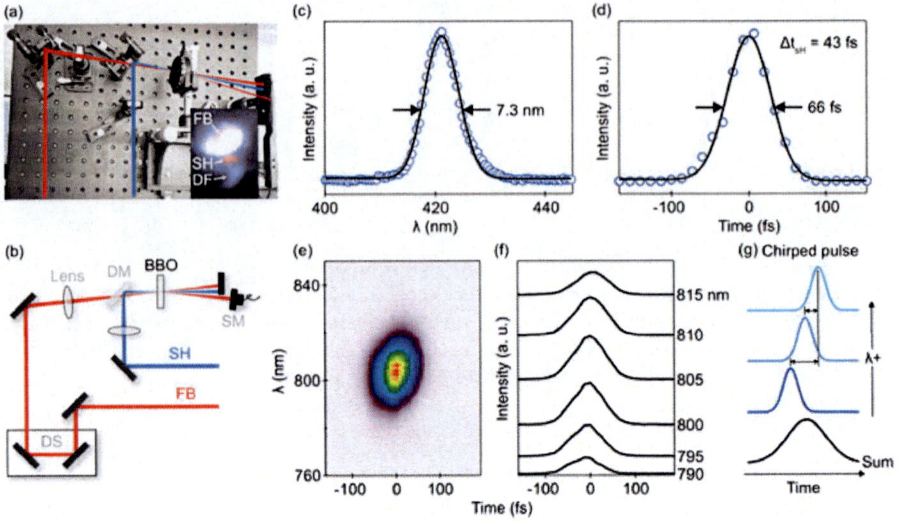

Fig. 3.7 Characterization of the SH. **a**, **b** Optical layout and a schematic of the SH diagnosis system. The insert shows beam spots of FB, SH, and DFG. **c** The spectrum of SH. **d** Cross-correlation from DFG between SH and FB. **e** DFG-FROG image from SH and FB. **f** Extracted time evolution curves at different wavelengths, indicating no chirp. **g** A schematic drawing of a chirped pulse in the time domain for different wavelengths. Reproduced, with permission from [48]

judged from the time shift of the cross-correlation curve at different wavelengths. The negligible time shift for different wavelengths as shown in Fig. 3.7f confirms that there is negligible chirp for the SH as expected.

3.2.3.4 Time Resolution Characterization Scheme

To evaluate the time resolution of the TrARPES setup, a reliable and efficient detection scheme is required. The time resolution can be extracted from the leading edge of the TrARPES signal, however, the leading edge is typically quite complicated due to multiple excitation channels of the photo-excited charge carriers, which are coupled to the lifetime of photo-excited carriers. Therefore, the choice of a good TrARPES sample is critical for efficient characterization and optimization of the time resolution.

We have tried polycrystalline gold, single crystal Bi_2Se_3 and molecular beam epitaxy (MBE) grown Sb_2Te_3 film on a bilayer graphene/SiC substrate. We find that the MBE grown Sb_2Te_3 film is the best for evaluating the TrARPES resolution with several advantages listed below. First of all, its TrARPES signal is much stronger than other samples, and a wide range of relaxation time scales depending on the choice of energy or momentum of the photoelectrons lead to ultra-high detection efficiency when searching for the TrARPES signal. For example, near the Fermi energy E_F, the relaxation is quite slow and photo-excited electrons above E_F persist up to 10

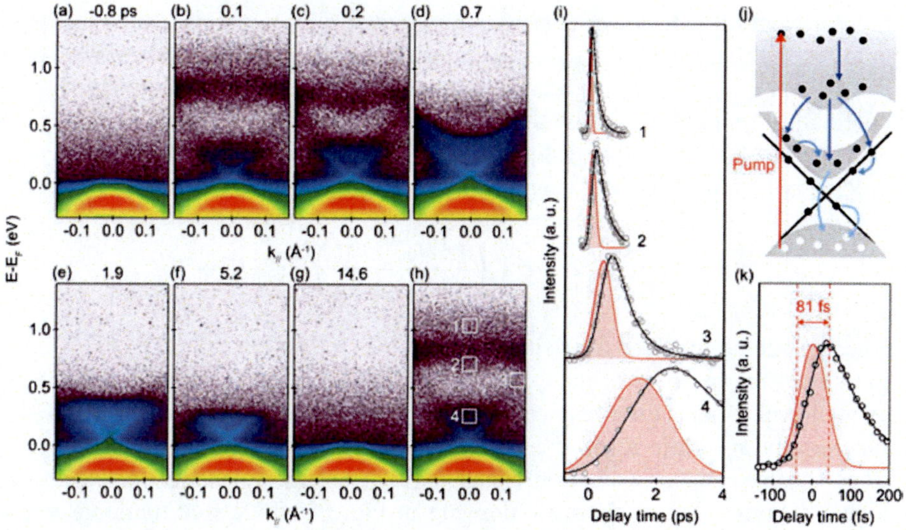

Fig. 3.8 Characterization scheme of the time resolution of the TrARPES system. **a–g** Snapshots of electronic dispersion at different delay times after photo-excitation. **h** Electronic dispersion at 0.1 ps after photo-excitation. **i** Integrated intensity as a function of delay time for regions indicated by boxes in (**h**). Red curves are the fitted gauss peaks for extracting the rising time. **j** Schematic drawing for photo-excitation and electronic relaxation. **k** The extracted time resolution from the intensity trace at the highest energy [box 1 in (**h**)]. The pump fluence is 10 uJ/cm^2. Reproduced, with permission from [48]

ps (corresponding to an optical path difference of 3 mm). This is an excellent energy window to look for the TrARPES signal and time zero point after adjusting the optical path. Thanks to the strong TrARPES signal, integrating the TrARPES signal for 1 s is enough to capture the photo-excited electrons and a searching speed of up to 2 mm/s is allowed due to their slow relaxation. Searching a delay range as large as 20 mm can be completed within 10 s even at a pump fluence as low as 10 μJ/cm^2. On the other hand, at high energy which corresponds to direct optical excitation, the relaxation is fast enough so that the measurement of the rising edge directly reflects the intrinsic time resolution of the TrARPES setup. Secondly, the MBE Sb_2Te_3 sample is very large, clean, and uniform with the size determined by the substrate (3 mm × 10 mm) which is desired for TrARPES. In contrast, typical single crystal or polycrystalline samples may have bad edges or defects, which might lead to a strong photoelectron signal by the pump excitation or interference of electronic trajectory due to the low kinetic energy. Thirdly, the average damage threshold is much higher and the sample still survives even when a large pump power of 5000 W W/cm^2 (corresponding to a pump fluence of 62 μJ/cm^2) is applied, allowing to use of high pump fluence even at a high repetition rate of 76 MHz.

Figure 3.8a–g shows snapshots of the electronic dispersions measured through the Γ point of Sb_2Te_3 film at different delay times. After pump excitation, electrons are excited to unoccupied states above the Fermi level (Fig. 3.8b). Electrons at the highest band relax to lower energy in hundreds of femtoseconds, and electrons in both the bulk and surface bands near the Fermi level relax much slower in several picoseconds. These different relaxation time scales are observed more clearly in the continuous temporal evolution at different selected energies as shown in Fig. 3.8i. The TrARPES signal in Fig. 3.8i is fitted by a Gaussian function convolved with the product of step function and single-exponential function [51]:

$$I(t) = A\left(1 + erf\left(\frac{t - t_0}{\Delta t} - \frac{\Delta t}{2\tau}\right)\right) e^{-\frac{t-t_0}{\tau}} + B, \tag{3.8}$$

where Δt is the width of the rising edge and τ is relaxation time. The extracted Δt gives the response time of the excitation. For bands at the highest energy, the only excitation channel is direct optical excitation and Δt reflects the intrinsic TrARPES time resolution, while at lower energies, e.g. region 2,3,4 marked in Fig. 3.8h, electrons can have many excitation channels, e.g. through relaxation of electrons at higher energies (see a schematic drawing in Fig. 3.8j) [52], and therefore a much longer rising edge is observed. Therefore, the intensity evolution at the highest band reflects direct optical excitation by the pump pulse and the intrinsic TrARPES time resolution.

3.2.3.5 Fast Optimization During the Daily Operation

The systematic optimization and diagnostics to achieve the highest time resolution would be tough work which takes a lot of time, especially considering changing the photon energy and all the conditions are changed. Therefore, in the daily operation, we need a fast scheme to achieve a relatively good time resolution as follows.

The first step is optimizing the laser to ensure enough bandwidth according to the uncertainty principle. The second step is optimizing the compressor for FB by getting the maximum SH power because a shorter pulse duration of FB leads to higher SHG efficiency and stronger SH power. For the same reason, the compressor for SH is optimized by maximizing FH power. Finally, we optimize the compressor for FH which is the most sensitive for the final time resolution of TrARPES. The compressor for FH is optimized by maximizing the TrARPES signal. First, we look for the maximum electron counts above the Fermi level by varying the delay time and recording the counts. Secondly, we adjust the compressor a little (for example moving the prism 1 mm) and look for the maximum electron counts by varying the delay time again, because the time zero point would be changed due to the adjustment of the compressor. Comparing the maximum counts with that before adjusting the compressor, if it increases, continue the movement of the prism, otherwise, move the prism in the opposite direction. Repeat these operations until the maximum electron counts are found.

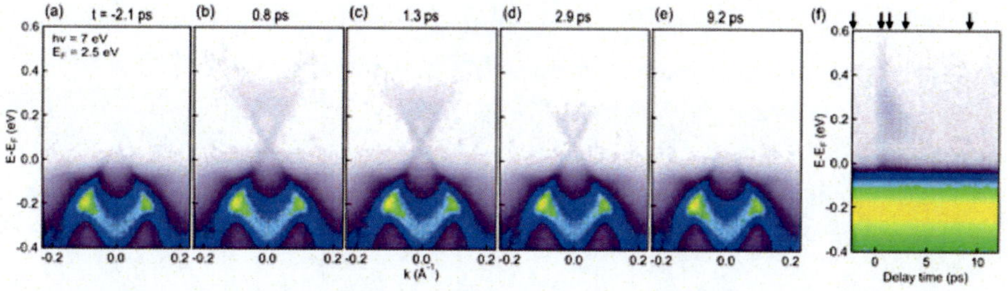

Fig. 3.9 The demonstration of TrARPES on a Sb_2Te_3 film. **a–e** Snapshots of electronic dispersion at different delay times as indicated by black arrows in (**f**) using probe photon energy of 7.0 eV (Fermi energy is 2.5 eV) and pump photon energy of 1.75 eV. **f** Corresponding integrated energy distribution curves as a function of delay time. The pump fluence is 80 μJ/cm^2 and the temperature is 80 K. Reproduced, with permission from [31]

3.2.4 TrARPES Functionality

The capability of the developed TrARPES system for capturing the ultrafast dynamics and the final time resolution is demonstrated on topological insulator Sb_2Te_3.

3.2.4.1 TrARPES on 3D Topological Insulator Sb_2Te_3

The sample is a thin film grown by molecular beam epitaxy (MBE) with a slight hole doping and only the dispersion below the Dirac point is observed at negative time delay (no pump) in Fig. 3.9a. Upon photoexcitation, electrons are excited to the unoccupied states above the Fermi energy, and the whole Dirac cone and bulk conduction band are observed in Fig. 3.9b. The photoexcited electrons relax to an equilibrium state in several picoseconds as shown in Fig. 3.9b–e. Furthermore, the evolution of photoexcited electrons is also measured by changing the delay time between the pump and probe laser pulses. As shown in Fig. 3.9f, the electronic states at high energy are dominated by direct photoexcitations and have a much shorter lifetime, which can be used to extract the time resolution.

3.2.4.2 Time Resolutions

The time resolution is obtained by analyzing the temporal evolution of photoexcited electrons at high energy measured on Sb_2Te_3 as described in the previous section. Figure 3.10a shows the images measured as a function of energy and delay time. By integrating over the energy range indicated by dashed lines in Fig. 3.10a, the intensity curve in Fig. 3.10b shows a fast rising edge. The time resolution (FWHM) is extracted to be 280 ± 20 fs for 5.9 eV. The time resolution is also extracted to be 280 ± 10 and 320 ± 80 fs for the probe photon energy of 6.4 and 7.0 eV respectively as shown in Fig. 3.10c–f. For 7.0 eV, a lower energy range is used to extract the time

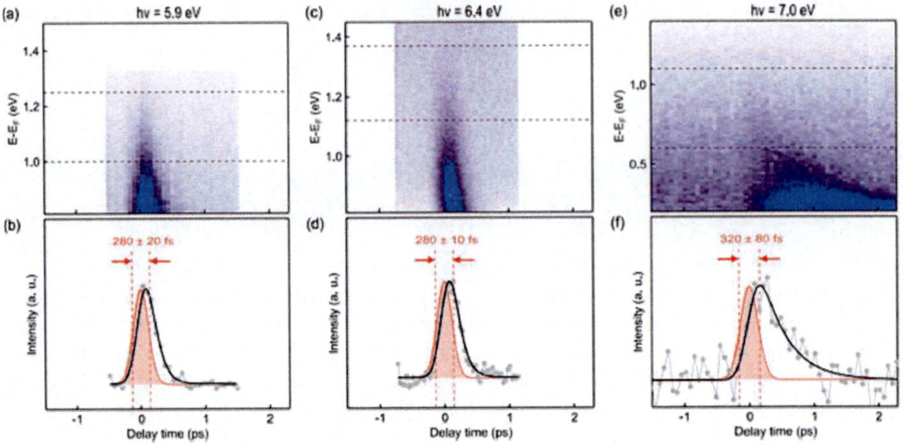

Fig. 3.10 The time resolution of the TrARPES system. **a**, **c**, **e** Integrated energy distribution curves as a function of delay time at probe photon energy of 5.9 (**a**), 6.4 (**c**) and 7.0 eV (**e**). The pump photon energies are 1.48, 1.60 and 1.75 eV for (**a**), (**c**), (**e**) respectively. **b**, **d**, **f** Integrated intensity as a function of delay time for regions between dashed lines in (**a**), (**c**) and (**e**) with a single-exponential relaxation fitting. Red curves are the fitted Gaussian peaks for extracting the time resolution. Reproduced, with permission from [31]

resolution due to the negligible excitation for the high-energy band. This explains the slower relaxation in Fig. 3.10f, and suggests that the real time resolution could be slightly better than the extracted value.

3.3 Conclusions

In summary, we develop a KBBF-based TrARPES system with a widely tunable probe photon energy VUV laser covering 5.3–7.0 eV. The energy resolution is 29–48 meV and the time resolution is 280–320 fs. The TrARPES system is demonstrated on 3D topological materials. Our work opens up new opportunities for exploring ultrafast light-matter interaction in 3D quantum materials [1, 14, 16, 53]. In the following chapters, I will present the study on both equilibrium and non-equilibrium electronic structure of Dirac semimetal by utilizing this developed TrARPES system.

References

1. Bao C, Tang P, Sun D et al (2022) Light-induced emergent phenomena in 2D materials and topological materials. Nat Rev Phys 4:33–48
2. Smallwood CL, Kaindl RA, Lanzara A (2016) Ultrafast angle-resolved photoemission spectroscopy of quantum materials. Europhys Lett 115(2):27001
3. Sobota JA, He Y, Shen ZX (2021) Angle-resolved photoemission studies of quantum materials. Rev Mod Phys 93:025006
4. Wang YH, Steinberg H, Jarillo-Herrero P et al (2013) Observation of Floquet-Bloch states on the surface of a topological insulator. Science 342(6157):453–457

5. Reimann J, Schlauderer S, Schmid C et al (2018) Subcycle observation of lightwave-driven Dirac currents in a topological surface band. Nature 562(7727):396–400
6. Faure J, Mauchain J, Papalazarou E et al (2012) Full characterization and optimization of a femtosecond ultraviolet laser source for time and angle-resolved photoemission on solid surfaces. Rev Sci Instrum 83(4):043109
7. Ishida Y, Togashi T, Yamamoto K et al (2014) Time-resolved photoemission apparatus achieving sub-20-mev energy resolution and high stability. Rev Sci Instrum 85(12):123904
8. Rohde G, Hendel A, Stange A et al (2016) Time-resolved ARPES with sub-15 fs temporal and near Fourier-limited spectral resolution. Rev Sci Instrum 87(10):103102
9. Gauthier A, Sobota JA, Gauthier N et al (2020) Tuning time and energy resolution in time-resolved photoemission spectroscopy with nonlinear crystals. J Appl Phys 128(9):093101
10. Mathias S, Miaja-Avila L, Murnane MM et al (2007) Angle-resolved photoemission spectroscopy with a femtosecond high harmonic light source using a two-dimensional imaging electron analyzer. Rev Sci Instrum 78(8):083105
11. Wang H, Xu Y, Ulonska S et al (2015) Bright high-repetition-rate source of narrowband extreme-ultraviolet harmonics beyond 22 ev. Nat Commun 6(1):7459
12. Sie EJ, Rohwer T, Lee C et al (2019) Time-resolved XUV ARPES with tunable 24–33 eV laser pulses at 30 meV resolution. Nat Commun 10(1):3535
13. Kutnyakhov D, Xian RP, Dendzik M et al (2020) Time- and momentum-resolved photoemission studies using time-of-flight momentum microscopy at a free-electron laser. Rev Sci Instrum 91(1):013109
14. Sun Z, Martinez A, Wang F (2016) Optical modulators with 2D layered materials. Nat Photonics 10(4):227–238
15. Chan CK, Oh YT, Han JH et al (2016) Type-II Weyl cone transitions in driven semimetals. Phys Rev B 94:121106
16. Hübener H, Sentef MA, Giovannini UD et al (2017) Creating stable Floquet-Weyl semimetals by laser-driving of 3D Dirac materials. Nat Commun 8:13940
17. Weber CP (2021) Ultrafast investigation and control of Dirac and Weyl semimetals. J Appl Phys 129(7):070901
18. Lv BQ, Qian T, Ding H (2021) Experimental perspective on three-dimensional topological semimetals. Rev Mod Phys 93:025002
19. Wu B, Tang D, Ye N et al (1996) Linear and nonlinear optical properties of the $KBe_2BO_3F_2$ (KBBF) crystal. Opt Mater 5(1–2):105–109
20. Chen C, Xu Z, Deng D et al (1996) The vacuum ultraviolet phase-matching characteristics of nonlinear optical $KBe_2BO_3F_2$ crystal. Appl Phys Lett 68(21):2930–2932
21. Chen C, Lu J, Togashi T et al (2002) Second-harmonic generation from a $KBe_2BO_3F_2$ crystal in the deep ultraviolet. Opt Lett 27(8):637–639
22. Zhang H, Wang G, Guo L et al (2008) 175 to 210 nm widely tunable deep-ultraviolet light generation based on KBBF crystal. Appl Phys B 93(2):323–326
23. McMillen CD, Kolis JW (2008) Hydrothermal crystal growth of $ABe_2BO_3F_2$ (A=K, Rb, Cs, Tl) NLO crystals. J Cryst Growth 310(7):2033–2038
24. Chen C, Wang G, Wang X et al (2009) Deep-UV nonlinear optical crystal $KBe_2BO_3F_2$-discovery, growth, optical properties and applications. Appl Phys B 97(1):9–25
25. Kiss T, Shimojima T, Ishizaka K et al (2008) A versatile system for ultrahigh resolution, low temperature, and polarization dependent Laser-angle-resolved photoemission spectroscopy. Rev Sci Instrum 79(2):023106
26. Liu G, Wang G, Zhu Y et al (2008) Development of a vacuum ultraviolet laser-based angle-resolved photoemission system with a superhigh energy resolution better than 1 meV. Rev Sci Instrum 79(2):023105
27. Jiang R, Mou D, Wu Y et al (2014) Tunable vacuum ultraviolet laser based spectrometer for angle resolved photoemission spectroscopy. Rev Sci Instrum 85(3):033902
28. Zhou X, He S, Liu G et al (2018) New developments in laser-based photoemission spectroscopy and its scientific applications: a key issues review. Rep Prog Phys 81(6):062101

29. Xu ZY, Zhang SJ, Zhou XJ et al (2019) Advances in deep ultraviolet laser based high-resolution photoemission spectroscopy. Front Inf Technol Electron Eng 20(7):885–913
30. Yang Y, Tang T, Duan S et al (2019) A time-and angle-resolved photoemission spectroscopy with probe photon energy up to 6.7 eV. Rev Sci Instrum 90(6):063905
31. Bao C, Zhong H, Zhou S et al (2022) Ultrafast time- and angle-resolved photoemission spectroscopy with widely tunable probe photon energy of 5.3–7.0 eV for investigating dynamics of three-dimensional materials[J/OL]. Rev Sci Instrum 93(1). https://doi.org/10.1063/5.0070004
32. Watanabe K, Inn EC, Zelikoff M (1953) Absorption coefficients of oxygen in the vacuum ultraviolet. J Chem Phys 21(6):1026–1030
33. Puppin M, Deng Y, Nicholson C et al (2019) Time-and angle-resolved photoemission spectroscopy of solids in the extreme ultraviolet at 500 kHz repetition rate. Rev Sci Instrum 90(2):023104
34. Zhang H, Bao C, Schuler M et al (2021) Self-energy dynamics and mode-specific phonon threshold effect in a Kekulé-ordered graphene. Natl Sci Rev 9(5):nwab175
35. Weng H, Dai X, Fang Z (2014) Transition-metal pentatelluride $ZrTe_5$ and $HfTe_5$: a paradigm for large-gap quantum spin Hall insulators. Phys Rev X 4(1):011002
36. Li Q, Kharzeev DE, Zhang C et al (2016) Chiral magnetic effect in $ZrTe_5$. Nat Phys 12(6):550–554
37. Moreschini L, Johannsen JC, Berger H et al (2016) Nature and topology of the low-energy states in $ZrTe_5$. Phys Rev B 94:081101
38. Manzoni G, Gragnaniello L, Autès G et al (2016) Evidence for a strong topological insulator phase in $ZrTe_5$. Phys Rev Lett 117:237601
39. Xiong H, Sobota JA, Yang SL et al (2017) Three-dimensional nature of the band structure of $ZrTe_5$ measured by high-momentum-resolution photoemission spectroscopy. Phys Rev B 95:195119
40. Zhang Y, Wang C, Yu L et al (2017) Electronic evidence of temperature-induced Lifshitz transition and topological nature in $ZrTe_5$. Nat Commun 8(1):1–9
41. Liu Z, Zhou B, Zhang Y et al (2014) Discovery of a three-dimensional topological Dirac semimetal, Na_3Bi. Science 343(6173):864–867
42. Ali MN, Gibson Q, Jeon S et al (2014) The crystal and electronic structures of Cd_3As_2, the three-dimensional electronic analogue of graphene. Inorg Chem 53(8):4062–4067
43. Lv BQ, Weng HM, Fu BB et al (2015) Experimental discovery of Weyl semimetal TaAs. Phys Rev X 5:031013
44. Averitt RD, Taylor AJ (2002) Ultrafast optical and far-infrared quasiparticle dynamics in correlated electron materials. J Phys Condens Matter 14(50):R1357–R1390
45. Perfetti L, Loukakos PA, Lisowski M et al (2008) Femtosecond dynamics of electronic states in the Mott insulator 1T-TaS_2 by time resolved photoelectron spectroscopy. New J Phys 10(5):053019
46. Rulliere C et al (2005) Femtosecond laser pulses. Springer
47. Trebino R (2000) Frequency-resolved optical gating. Kluwer Academic Publishers
48. Bao C, Luo L, Zhang H et al (2021) Full diagnostics and optimization of time resolution for time- and angle-resolved photoemission spectroscopy. Rev Sci Instrum 92 (3)
49. Fork RL, Martinez OE, Gordon JP (1984) Negative dispersion using pairs of prisms. Optics Lett 9(5):150
50. Trebino R, Kane DJ (1993) Using phase retrieval to measure the intensity and phase of ultrashort pulses: frequency-resolved optical gating. J Opt Soc Am A 10(5):1101–1111
51. Smallwood CL, Hinton JP, Jozwiak C et al (2012) Tracking cooper pairs in a cuprate superconductor by ultrafast angle-resolved photoemission. Science 336:1137
52. Sobota JA, Yang S, Analytis JG et al (2012) Ultrafast optical excitation of a persistent surface-state population in the topological insulator Bi_2Se_3. Phys Rev Lett 108:117403
53. de la Torre A, Kennes DM, Claassen M et al (2021) Colloquium: nonthermal pathways to ultrafast control in quantum materials. Rev Mod Phys 93:041002

Chapter 4
Chiral Symmetry Breaking in Kekulé-Ordered Graphene

Check for updates

4.1 Research Motivation

Chirality $\hat{h}$ is a fundamental property for relativistic massless Dirac fermions, which is defined by the projection of spin $\boldsymbol{\sigma}$ onto momentum $\hat{\boldsymbol{p}}$, namely, $\hat{h} = \frac{1}{2}\boldsymbol{\sigma} \cdot \hat{\boldsymbol{p}}/|\hat{\boldsymbol{p}}|$. Chiral symmetry breaking (CSB), namely, coupling of Dirac fermions with opposite chiralities, leads to dynamical mass generation for elementary particles [1], which lays a cornerstone for the Standard Model in particle physics. The low-energy excitations of graphene are massless Dirac fermions [2, 3] with the electron spin replaced by the pseudospin, and therefore graphene provides a condensed matter physics analogue for investigating CSB and mass generation which shows up band gap opening [4–9]. In addition to gap opening, a series of intriguing phenomena have been proposed with CSB, such as electron fractionalization [10] and topological effects in the electron and phonon spectra [11, 12]. Experimental realization and unambiguous observation of CSB in graphene are therefore highly desirable.

In order to realize CSB, a superlattice period of $(\sqrt{3} \times \sqrt{3})R30°$ is required to couple Dirac cones from K and K′ valleys, and such inter-valley coupling could lead to replica Dirac cones at the Brillouin zone (BZ) center and gap opening as schematically illustrated in Fig. 4.1a. Such superlattice period is inherent in the Kekulé distortion with modulated bond structure as indicated by red and black bonds in Fig. 4.1b. Experimentally, Kekulé distortion can be induced by a strong magnetic field [14–16], which however also breaks the time-reversal symmetry and turns the conical dispersion into Landau levels. Introducing dilute adatoms or defects [5, 17–19] or an external superlattice potential [20–22] provides an attractive pathway for realizing CSB while still preserving the conical dispersion. Along this path, important progress has been made in graphene grown on Cu substrate, where Kekulé-Y order with "Y"-shaped bond modulation as shown in Fig. 4.1c has been revealed by scanning tunneling microscopy (STM) [17]. However, there is no evidence of gap opening, and recent theoretical calculation suggests that such Kekulé-Y ordered graphene would remain gapless, while Kekulé-O order where modulated bonds form an "O"-shape pattern as shown in Fig. 4.1d is gapped [18] which has been realized

© Tsinghua University Press 2023

C. Bao, *Electronic Band Structure Engineering and Ultrafast Dynamics of Dirac Semimetals*, Springer Theses, https://doi.org/10.1007/978-981-99-5325-7_4

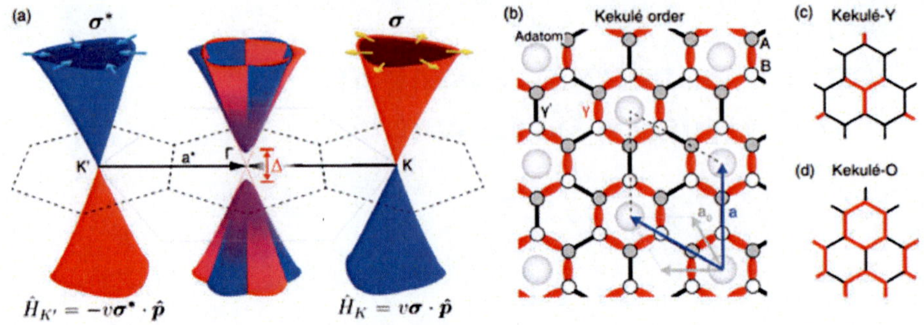

Fig. 4.1 Chiral symmetry breaking in Kekulé-ordered graphene. **a** A schematic for CSB. Straight blue and yellow arrows point to the pseudospin directions. Blue and yellow circular arrows near the Dirac points indicate negative and positive chiralities. **b** A Kekulé-ordered graphene with different bond strengths (represented by black and red colors) and nearest-neighbor hopping parameters γ and γ'. The unit cells of pristine and Kekulé-ordered graphene are marked by gray and blue parallelograms. **c** Kekulé-Y bond texture. **d** Kekulé-O bond texture. Reproduced, with permission from [13]

in artificial graphene-like structure. Superlattice period of $(\sqrt{3} \times \sqrt{3})R30°$ has also been reported in Li or Ca intercalated graphene [23–29] or graphite [30, 31], making them potential candidates for realizing CSB. However, experimental evidence such as CSB induced gap opening, a specific type of Kekulé order, and chirality mixing near the gap edge, which are crucial for establishing the CSB, are still missing.

4.2 Experimental Evidence of Chiral Symmetry Breaking

We look for the evidence of chiral symmetry breaking in Li-intercalated graphene by combing both microscopic and spectroscopic measurements.

4.2.1 Preparation of Kekulé-Ordered Graphene

Kekulé-ordered graphene is prepared by intercalating Li atoms to monolayer graphene on SiC substrate, so a high-quality graphene sample is important.

4.2.1.1 High-Quality Graphene Sample

High-quality graphene with uniform thickness was grown by flash annealing the Si face of 6H-SiC(0001) substrates in ultrahigh vacuum [32]. Figure 4.2a, b shows the atomic force microscopy (AFM) image of the monolayer graphene sample (one

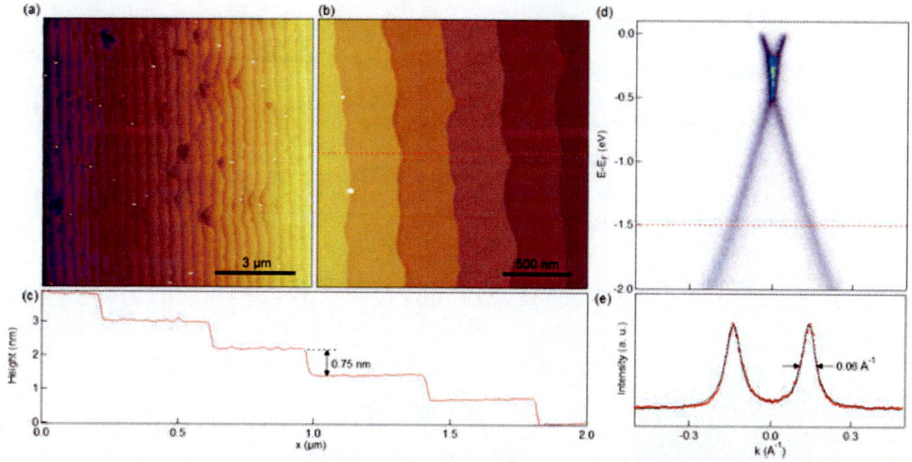

Fig. 4.2 High-quality graphene sample. **a, b** Atomic force microscopy images of the monolayer graphene sample used for the Li intercalation. **c** The line profile along the dashed line in (**b**). **d** ARPES dispersion image of monolayer graphene. **e** Momentum distribution curve at the energy of −1.5 eV. Reproduced, with permission from [13]

graphene layer + one buffer layer). A uniform step height of 0.75 Å corresponding to the triple SiC bilayer step [32] is shown in Fig. 4.2c, which suggests that the SiC substrate is covered with uniform graphene thickness. The high sample quality is confirmed by the sharp ARPES dispersion shown in Fig. 4.2d, which shows no detectable contribution from bilayer or trilayer graphene. The ARPES peak width is also another indicator for the high-quality graphene sample. In our experiments, the best graphene sample with a peak width of better than 0.06 Å^{-1} is used for the intercalation. Such a high-quality graphene sample is obtained by an automatically controlled growth procedure and improving the stability of the growth temperature and is critical for obtaining the high-quality Li-intercalated graphene.

4.2.1.2 Li-Intercalated Graphene

The Kekulé order is introduced by intercalating Li [23] to monolayer graphene on SiC substrate as shown in Fig. 4.3. Li intercalation was performed by in situ deposition of Li using an alkali metal dispenser (SAES) [23], with the graphene samples maintained at 320 K. The intercalation process was monitored by LEED and ARPES measurements. The intercalation releases the bonding between the buffer layer [33] and the SiC substrate, which transforms the buffer layer into a new graphene layer [23, 28] and results in two graphene layers in the AA stacking [29]. We note that intercalation of the buffer layer does not lead to Kekulé order [28, 34], and two graphene layer structure has the minimum thickness required to stabilize the Kekulé order.

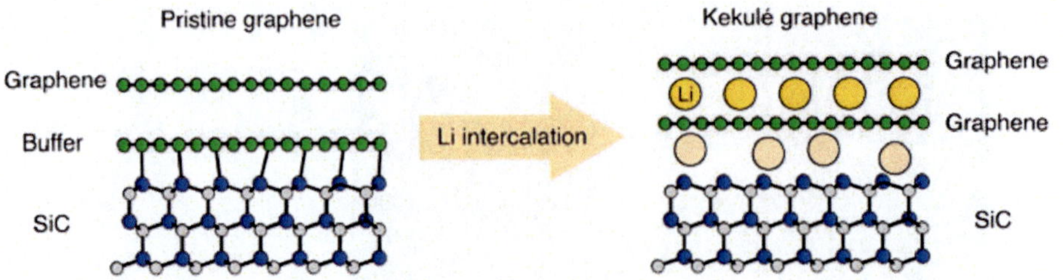

Fig. 4.3 Schematics for preparation of Kekulé graphene sample. Reproduced, with permission from [13]

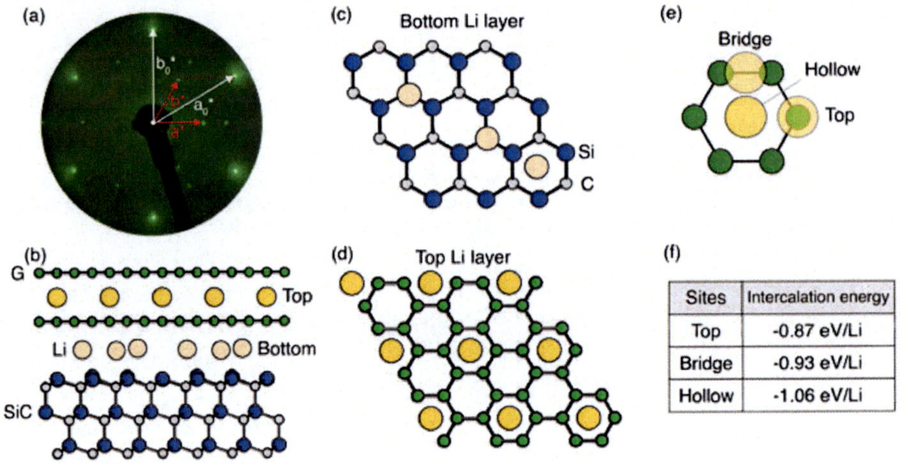

Fig. 4.4 Atomic structures of Li-intercalated graphene. **a** LEED pattern for Li-intercalated graphene. White and red arrows are the unit vectors for graphene and Kekulé order. **b** Atomic structures of the Li-intercalated graphene sample. **c, d** Li configurations for the bottom and top intercalated Li layers. **e** Three possible Li sites for the top Li layer with respect to the graphene sheet: top, hollow, and bridge. **f** Corresponding intercalation energy for the three different sites for the top Li layer. Reproduced, with permission from [13]

Density functional theory (DFT) calculations have been performed to investigate the intercalation of Li. When Li atoms are intercalated between the buffer layer and the SiC substrate, they prefer to locate at the hollow site of terminated Si atoms of the SiC substrate due to the dangling bonds of Si atoms as shown in Fig. 4.4c. Such intercalation releases the bonding between the buffer layer and the SiC substrate, however, intercalation of the buffer layer does not lead to $(\sqrt{3} \times \sqrt{3})R30°$ order [28, 34]. For Li atoms intercalated between the two graphene layers, they prefer to locate at the hollow site of graphene (Fig. 4.4e, f), resulting in two intercalated graphene layers with AA stacking and $(\sqrt{3} \times \sqrt{3})R30°$ order (Fig. 4.4a, d) [27, 29]. To investigate the stability of Li intercalation, the deintercalation energy of Li atoms has also been calculated, which are 2.23 eV/Li and 2.73 eV/Li for the top and bottom Li layers, respectively. Such deintercalation energy is much larger

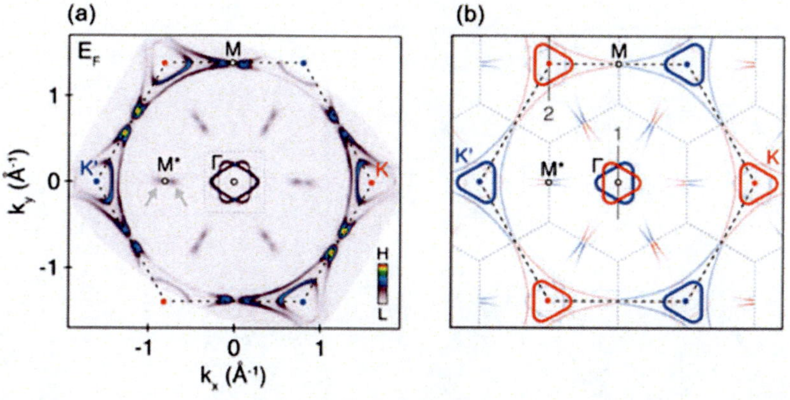

Fig. 4.5 Kekulé order induced band folding. **a** Experimental Fermi surface map of the Kekulé-ordered graphene measured using helium lamp source at 21.2 eV. The intensity inside the dashed box around Γ is enhanced for better visualization. **b** A schematic Fermi surface map of the Kekulé-ordered graphene. Reproduced, with permission from [13]

than the desorption energy of Li from the outer surface of the top graphene layer (1.41 eV/Li), indicating that Li atoms prefer to intercalate between the two graphene layers or between graphene and SiC substrate rather than adsorption above the top graphene layer. In summary, the $(\sqrt{3} \times \sqrt{3})R30°$ Kekulé order develops through Li atoms intercalation between the two graphene layers.

4.2.1.3 Band Folding Due to the Kekulé Order

After intercalation, the Fermi surface map in Fig. 4.5a shows two trigonal pockets at each BZ corner, a hexagonal star-shaped pocket at the Γ point, and arcs near the M* point. The small and large triangular pockets originate from the top and bottom graphene layers respectively. The star-shaped pocket at Γ is formed by the superposition of the folded inner triangular pockets from K and K′ by the reciprocal Kekulé superlattice vector (see schematic drawing in Fig. 4.5b). Replicas of the larger triangular pockets folded from K and K′ to Γ, and K to K′ or vice versa are also observed near the M* point (pointed by gray arrows in Fig. 4.5a). The existence of folded Dirac cones by $(\sqrt{3} \times \sqrt{3})R30°$ superlattice modulation [24, 25, 31, 35] is a necessary but insufficient condition for CSB, and further experimental evidence is needed to confirm the CSB, which is the main focus of this work.

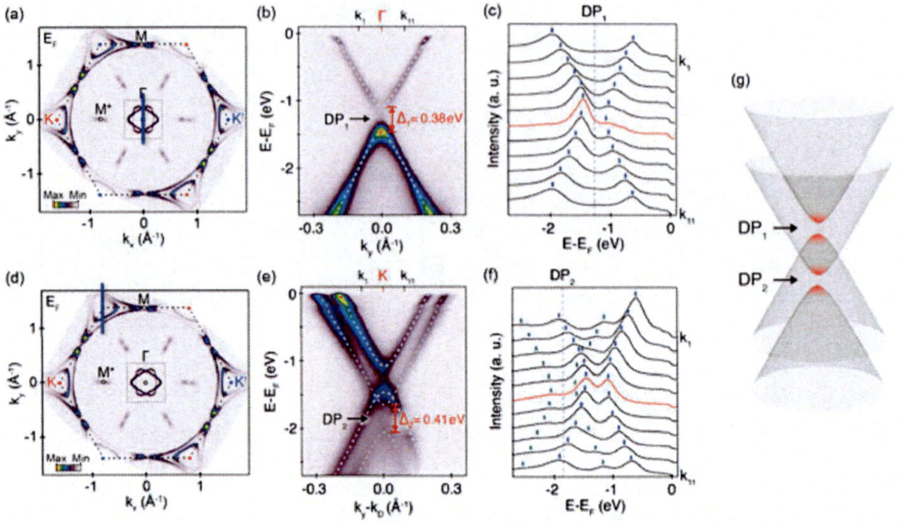

Fig. 4.6 Band gap opening at the Dirac points. **a, d** Experimental Fermi surface map of the Kekulé-ordered graphene and the blue lines indicate the cut directions for dispersion image in (**b**, **e**). **b** Dispersion image of the folded Dirac cones measured along the M-Γ-M direction [marked by the blue line in (**a**)]. **c** EDCs for data shown in (**b**). **e** Dispersion image of Dirac cones through the K point and perpendicular to K-M direction [marked by the blue line in (**a**)]. **f** EDCs for data shown in (**e**). **g** A schematic summary of two gapped Dirac cones as observed experimentally. Reproduced, with permission from [13]

4.2.2 *Evidence of CSB 1: Band Gap Opening*

An important evidence of CSB is the gap opening at the Dirac point, which is revealed by ARPES dispersions measured through both Γ and K points.

4.2.2.1 Band Gap Opening at the Dirac Point

A gap of 380 ± 10 meV is clearly identified for the upper Dirac cone (inner pocket) from the dispersion image shown in Fig. 4.6b as well as energy distribution curves (EDCs) shown in Fig. 4.6c. Such gap value corresponds to an effective mass of $m^* = 0.033\, m_e$, where m_e is the free electron mass. The gap opening is also observed in the dispersion image measured along an equivalent cut through the K point (Fig. 4.6e). Here, not only the upper Dirac cone but also the lower Dirac cone with a larger pocket is clearly resolved. The upper Dirac cone shows a gap value consistent with Fig. 4.6c with the Dirac point at −1.27 eV (broken line in Fig. 4.6c), and the lower Dirac cone shows a gap of 410 ± 50 meV ($m^* = 0.036\, m_e$) with the Dirac point at −1.82 eV (broken line in Fig. 4.6f). Our ARPES data show that these two Dirac cones are both gapped and shifted in energy, as schematically summarized in Fig. 4.6g.

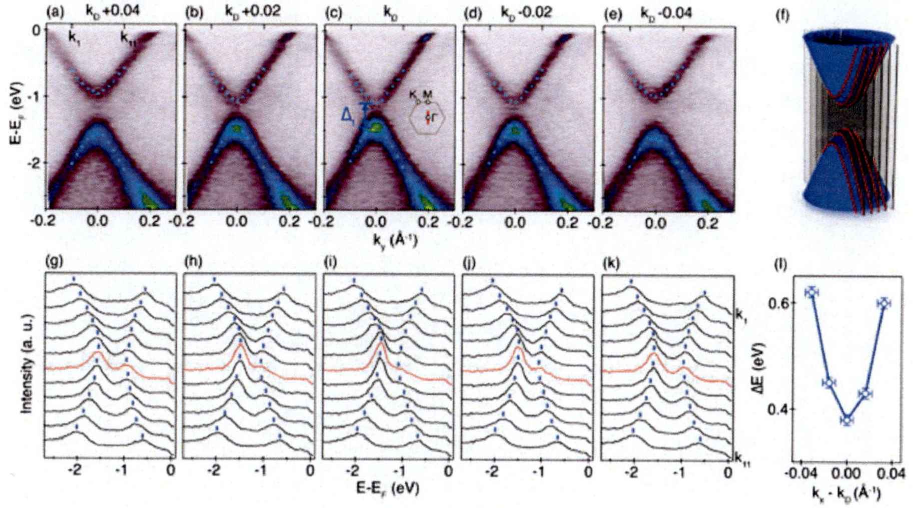

Fig. 4.7 Confirmation of the gap opening at the Γ point. **a–e** Dispersions measured along directions parallel to Γ-M at $k_x - k_D = 0.04, 0.02, 0, -0.02$ and -0.04 Å^{-1} [see schematic in (**f**)], respectively. Dotted curves are extracted dispersions from EDCs in (**g–k**). **f** Schematic drawing of the geometry for cuts shown in (**a–e**). **g–k** Corresponding EDCs for data in (**a–e**) at momentum from k_1 to k_{11}. Tick marks indicate the peak positions with an error bar of 10 meV. **l** Summary of the energy separation between the bottom of the conduction band and the top of the valence band for data from (**a–e**). Reproduced, with permission from [13]

4.2.2.2 Evolution of Dispersions Near the Γ and K Points

To confirm that the gap opening is a real effect and not due to misalignment from the Dirac point, we show in Figs. 4.7 and 4.8 a series of cuts through the Γ and K points. The extracted dispersions from corresponding EDCs are overplotted in the dispersion images, and the energy separation between the conduction and valence bands are summarized in Figs. 4.7l and 4.8l, respectively. The minimum separations of 380 and 410 meV define the gap values for the two Dirac cones.

4.2.2.3 Band Structures from Density Functional Theory Calculations

On the basis of the above Li configuration (Fig. 4.4), the electronic structure is calculated and compared with the experimental results. With both top and bottom intercalated Li layers (Fig. 4.9a), a gap opening is observed for both Dirac cones (Fig. 4.9d), consistent with our experimental results. In particular, the upper Dirac cone shows a gap of 0.17 eV, while the lower Dirac cone shows a gap opening of 0.45 eV, in qualitative agreement with the experimental results. To further elucidate which intercalated Li layer induces the gap, the electronic structures with only the top or bottom Li layers are investigated separately. As shown in Fig. 4.9b, e, when the top Li layer is intercalated between the two graphene layers resulting in Kekulé order

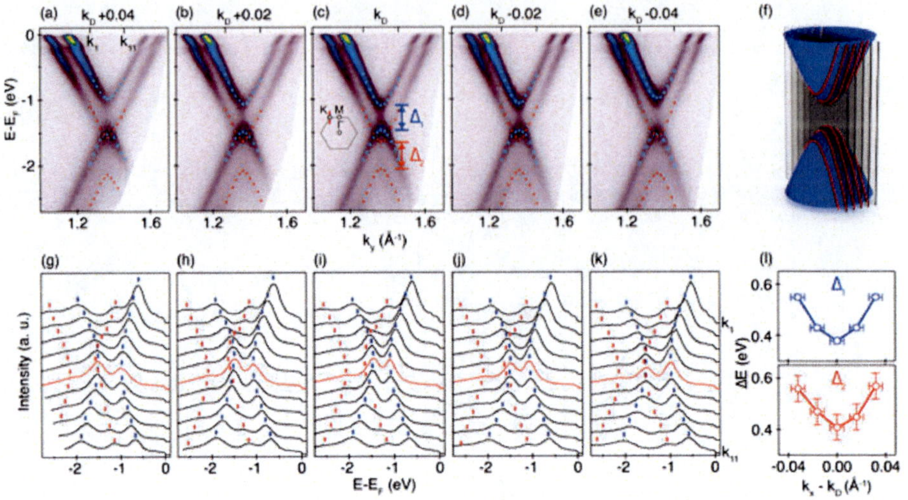

Fig. 4.8 Confirmation of the gap opening at the K point. **a–e** Dispersions measured along directions perpendicularly to the K-M direction at k_x-k_D = 0.04, 0.02, 0, −0.02 and −0.04 Å^{-1} [see schematic in (**f**)], respectively. Dotted curves are extracted dispersions from EDCs in (**g–k**). **f** Schematic drawing of the geometry for cuts shown in (**a–e**). **g–k** Corresponding EDCs for data in (**a–e**) at momentum from k_1 to k_{11}. Tick marks indicate the peak positions. The error bar is 10 meV for blue tick marks and 50 meV for red tick marks due to the broader peaks. **l** Summary of the energy separation between the bottom of the conduction band and the top of the valence band for data from (**a–e**). Reproduced, with permission from [13]

while keeping the bottom graphene layer decoupled from the SiC substrate, the two Dirac cone remains gapped with a gap size of 0.16 eV for the upper Dirac cone and 0.40 eV for the lower Dirac cone. However, the intercalation of the bottom Li layer can only induce a small gap opening for the lower Dirac cone and a negligible gap for the upper Dirac cone as shown in Fig. 4.9c, f. The comparison shows that the Kekulé order induced by the intercalated Li layer between the two graphene layers plays a dominant role in the gap opening.

4.2.3 Evidence of CSB 2: Kekulé-O Order

As noted above, recent theoretical calculation shows that only a certain type of Kekulé order can generate a gap opening [18], and therefore it is important to identify the specific type of Kekulé order. Here we directly probe the spatially-resolved electronic properties by performing STM topographic measurements at different bias voltages. Away from the Dirac point energy, for example, at −0.1 eV (Fig. 4.10a), the STM topography shows a honeycomb structure which is typical of graphene, while at energies near the gap regions of the two Dirac cones (Fig. 4.10b–e), the STM topography clearly demonstrates strongly modified topography, with a periodic arrangement of expanded hexagons (labeled by thicker blue bonds) alternating

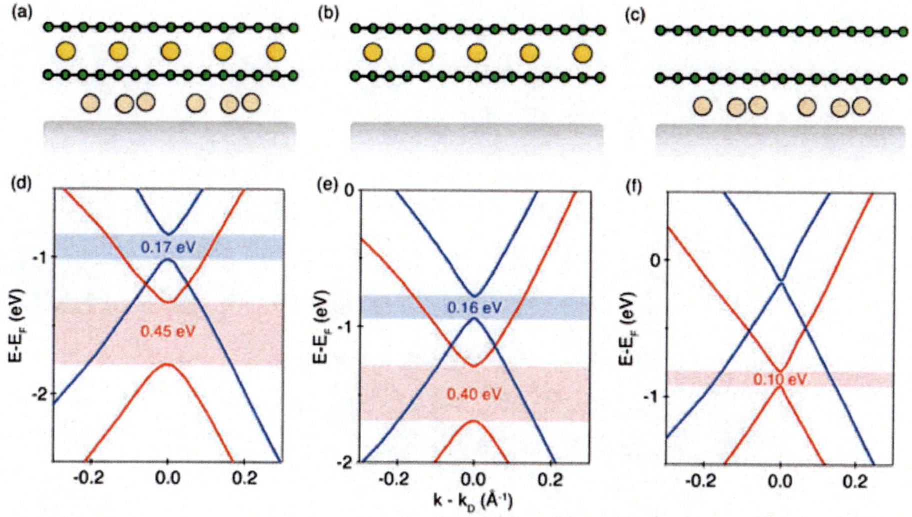

Fig. 4.9 Electronic structures of the graphene sample with different Li-intercalated layers from density functional theory calculations. **a–c** Atomic structures for the graphene sample with two Li-intercalated layers (**a**), only the top (**b**) and bottom (**c**) Li-intercalated layer. **d–f** Projected band structures of the graphene along Γ-K-M direction for different Li configurations as shown in (**a–c**). Reproduced, with permission from [13]

with highly distorted hexagons compatible with the Kekulé-O type pattern. Such Kekulé-O order is critical for inducing CSB gap [18]. Compared to other decorated graphene [17, 25, 26, 28] where $(\sqrt{3} \times \sqrt{3})R30°$ Kekulé order is also observed in the fast Fourier transformation image (inset of Fig. 4.10c), our work is the first to demonstrate Kekulé-O type modulation and to show that it exists only near the gap edge of the Dirac cone as schematically summarized in Fig. 4.10f, by combining the advantages of both STM and ARPES measurements.

The fact that the Kekulé-O pattern is observed only near the gap edge indicates that it is a modulation of the electronic states rather than a simple modulation of the lattice which is energy independent. This suggests that although the CSB has a somewhat different origin (induced by an external superlattice potential imposed by the intercalated Li atoms rather than a spontaneous Coulomb interaction in quantum electrodynamics [36]), the results of the CSB, including CSB induced gap opening and modulated electronic states near the gap edge, remain the same.

4.2.4 Evidence of CSB 3: Chirality Mixing

Mixing of opposite chiralities is important to probe the chiral symmetry breaking. We make one step further by pursuing the evolution of the chirality mixing near the gap edge through polarization-dependent ARPES measurements.

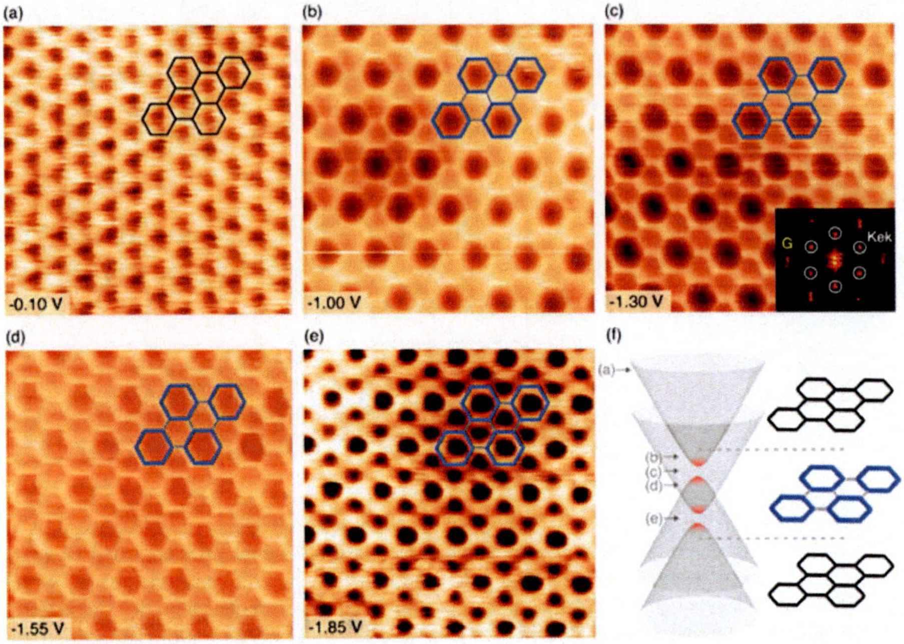

Fig. 4.10 Observation of Kekulé-O order in real space. **a–e** STM topographic images at different bias voltages. The size of all images is 2.5 nm×2.5 nm and the tunneling current is 0.5 nA (**a**) and 4.1 nA (**b–e**). The inset in (**c**) is two-dimensional fast Fourier transformation image of data in (**c**). **f** A schematic summary of Kekulé-O type modulation which is observed only near the gap edge. Reproduced, with permission from [13]

4.2.4.1 Method for Extraction of Pseudospin and Chirality

The accumulated pseudospin angle around a closed-loop (the Berry phase of π or 2π) can be deduced from the angular dependence of ARPES intensity using different light polarizations [37–39] as shown in Fig. 4.11a. Here we extend this experimental method for extracting the energy- and momentum-resolved pseudospin and chirality.

The ARPES intensity is proportional to $|M|^2 = |\langle f|\boldsymbol{A}\cdot\hat{\boldsymbol{p}}|\psi\rangle|^2$, where $\boldsymbol{A}$ is the vector potential of the incident light, $\hat{\boldsymbol{p}}$ is the momentum operator, $|\psi\rangle$ is the initial state, and $|f\rangle$ is the final state. For K valley, the initial state can be written as $|\psi\rangle = |A\rangle + e^{i\theta}|B\rangle$, where $|A\rangle$ and $|B\rangle$ are electronic states on two carbon sublattices, and θ determines the pseudospin by $\boldsymbol{\sigma} = (\cos\theta, \sin\theta)$ [2]. From a straightforward calculation [39], the ARPES intensity can be deduced to

$$I \propto |M|^2 = 2|\xi_x|^2 A_0^2[1 + \cos(\theta + 2\chi)], \tag{4.1}$$

where A_0 is the amplitude of vector potential, $\chi = \arg(A_x + A_y)$ is the phase angle of the light polarization and ξ_x the dipole transition matrix elements for the carbon p_z orbital excited by the x component of the vector potential. Therefore, the pseudospin θ is encoded in the ARPES intensity and coupled to χ.

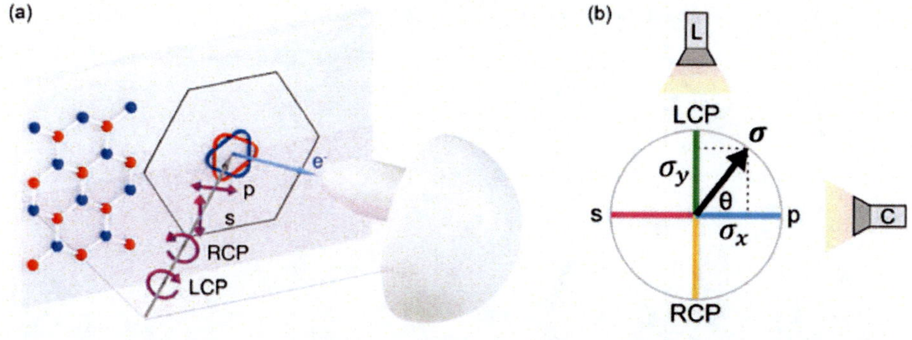

Fig. 4.11 Measurement of pseudospin and chirality by polarization-dependent ARPES. **a** Experimental geometry for the polarization-dependent ARPES measurements. **b** A schematic for probing the pseudospin σ by light polarization. Linear and circular polarizations probe the x (σ_x) and y (σ_y) components of the pseudospin σ respectively. Reproduced, with permission from [13]

For horizontal (p) and vertical (s) linear polarizations, $\chi = 0$ and π and the ARPES intensity becomes [39].

$$I_{p,s} \propto 2|\xi_x|^2 A_0^2(1 \pm \cos\theta) \tag{4.2}$$

where the $+$ and $-$ signs correspond to p and s respectively. Therefore, the linear dichroism can be used to extract the x component of the pseudospin $\sigma_x = \cos\theta$ directly by

$$I_{LD} = \frac{I_p - I_s}{I_p + I_s} = \cos\theta = \sigma_x. \tag{4.3}$$

Similarly, for left and right circular polarization, $\chi = \pm\frac{\pi}{4}$ and the ARPES intensity becomes [39]

$$I_{LCP,RCP} \propto 2|\xi_x|^2 A_0^2(1 \pm \sin\theta) \tag{4.4}$$

where the $+$ and $-$ signs correspond to LCP and RCP respectively. The circular dichroism therefore gives the y component of the pseudospin

$$I_{CD} = \frac{I_{LCP} - I_{RCP}}{I_{LCP} + I_{RCP}} = \sin\theta = \sigma_y. \tag{4.5}$$

Since the electron momentum is also directly measured by ARPES, we can calculate $\frac{1}{2}\boldsymbol{\sigma} \cdot \frac{\hat{p}}{|\hat{p}|}$ and extract the chirality.

The pseudospin for K′ valley can be easily obtained by time-reversal operation, because K′ valley is related by K valley through time-reversal symmetry (TRS) [2]. The linear polarization is unchanged under time-reversal operation, so the LD-ARPES is the same at K′ valley. In contrast, LCP and RCP are switched under time reversal operation, so $I_{CD}(K') = -I_{CD}(K)$ [40]. Therefore,

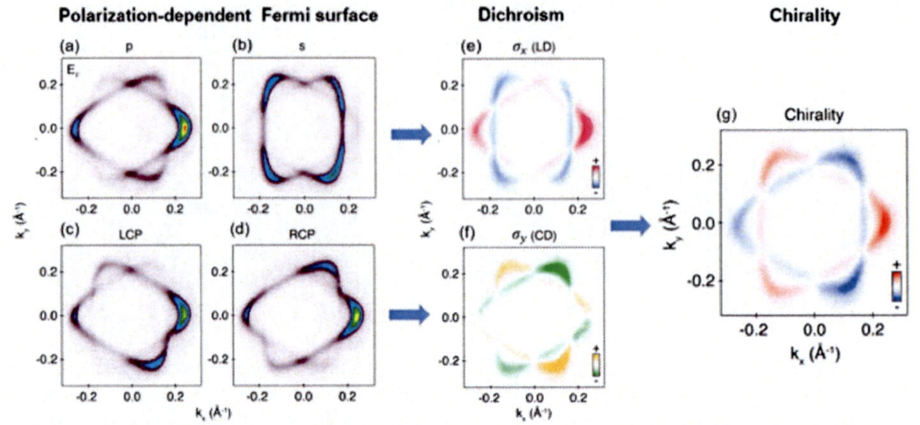

Fig. 4.12 Chirality-resolved Fermi surface. **a–d** Experimental Fermi surface maps measured around the Γ point using 11 eV laser source with four different light polarizations. **e, f** Extracted σ_x and σ_y from data in (**a–d**). **e, f** Experimental σ_x and σ_y maps at E_F probed by linear and circular polarized light. **g** Extracted chirality maps at E_F. Reproduced, with permission from [13]

$$I_{LD} = \cos\theta = \sigma_x^*, \tag{4.6}$$

$$I_{CD} = -\sin\theta = \sigma_y^*. \tag{4.7}$$

Therefore, for K′ valley, LD- and CD-ARPES measurements provide direct information on σ^* as summarized in Fig. 4.11b.

4.2.4.2 Chirality Mixing Near the Gap Edge

Figure 4.12a–d shows Fermi surface maps measured using four different light polarizations. The extracted momentum-resolved σ_x and σ_y maps from Fig. 4.12a–d are shown in Fig. 4.12e, f. By further taking the projection $\hat{h} = \frac{1}{2}\boldsymbol{\sigma} \cdot \hat{\boldsymbol{p}}/|\hat{\boldsymbol{p}}|$, opposite chiralities (represented by red and blue colors) are observed for folded Dirac cones from K and K′ at E_F (Fig. 4.12g), endorsing the validity of this analysis method.

Moreover, chirality maps measured at different energies (Fig. 4.13a) and chirality-resolved dispersion image along the Γ-K direction (Fig. 4.13b) show that although a strong chirality contrast is observed away from the Dirac point (e.g., at E_F and −2.4 eV), the chirality contrast becomes negligible in regions where folded Dirac cones from K and K′ (blue and red curves) overlap (pointed by black arrows in Fig. 4.13b), indicating that there is no well-defined chirality near the gap edge of the upper Dirac cone, which shows that the chirality contrast decreases gradually when moving toward the Dirac point and becomes negligible near the gap edge. Such energy-dependent chirality made available by polarization-dependent ARPES measurements confirms the chirality mixing near the gap edge, providing another important experimental evidence for CSB.

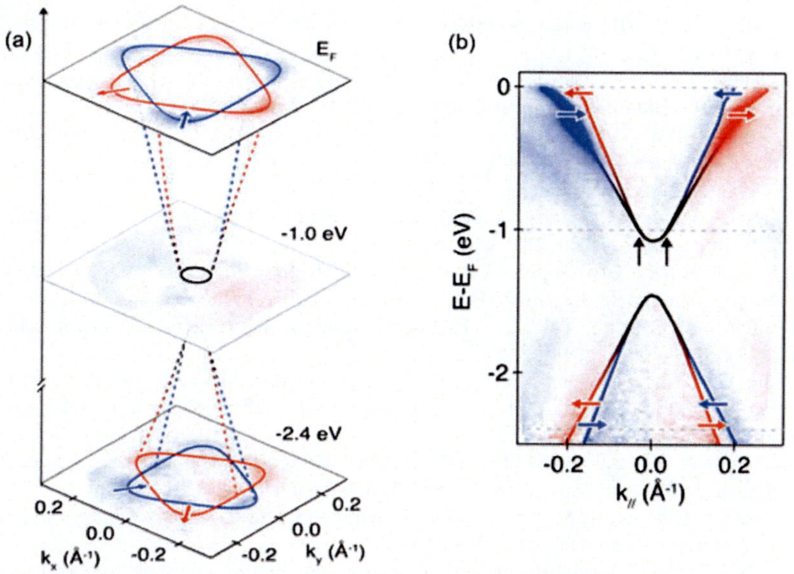

Fig. 4.13 Chirality mixing near the gap edge. **a** Extracted chirality maps at E_F, -1.0 and -2.4 eV. Red, blue and black curves are guides for the pockets, and red and blue arrows indicate the pseudospin directions. **b** Chirality-resolved dispersion image. Black arrows point to chirality mixing near the gap edge. Reproduced, with permission from [13]

4.3 Conclusion

In summary, the observation of replica Dirac cones with CSB induced gap opening, Kekulé-O patterned topography, and chirality mixing near the gap edge together provides definitive experimental evidences for CSB in a condensed matter physics system, in analogy to the mass generation in particle physics. Moreover, the experimental realization of CSB in the Kekulé-ordered graphene provides new opportunities for exploring CSB related physics, such as electron fractionalization [10] and topological effect [11, 12].

References

1. Nambu Y, Jona-Lasinio G (1961) Dynamical model of elementary particles based on an analogy with superconductivity. I. Phys Rev 122:345–358
2. Castro Neto AH, Guinea F, Peres NMR et al (2009) The electronic properties of graphene. Rev Mod Phys 81:109–162
3. Kotov VN, Uchoa B, Pereira VM et al (2012) Electron-electron interactions in graphene: current status and perspectives. Rev Mod Phys 84:1067–1125
4. Chamon C (2000) Solitons in carbon nanotubes. Phys Rev B 62:2806–2812
5. Cheianov VV, Fal'ko VI, Syljuåsen O et al (2009) Hidden Kekulé ordering of adatoms on graphene. Solid State Commun 149(37):1499–1501

6. González J (2010) Renormalization group approach to chiral symmetry breaking in graphene. Phys Rev B 82:155404
7. Lee SH, Chung HJ, Heo J et al (2011) Band gap opening by two-dimensional manifestation of peierls instability in graphene. ACS Nano 5(4):2964–2969
8. Semenoff GW (2012) Chiral symmetry breaking in graphene. Phys Scr 146:014016
9. Roy B, Kennett MP, Das Sarma S (2014) Chiral symmetry breaking and the quantum Hall effect in monolayer graphene. Phys Rev B 90:201409
10. Hou CY, Chamon C, Mudry C (2007) Electron fractionalization in two-dimensional graphene-like structures. Phys Rev Lett 98:186809
11. Wu LH, Hu X (2016) Topological properties of electrons in honeycomb lattice with detuned hopping energy. Sci Rep 6:24347
12. Liu Y, Lian CS, Li Y et al (2017) Pseudospins and topological effects of phonons in a Kekulé lattice. Phys Rev Lett 119:255901
13. Bao C, Zhang H, Zhang T et al (2021) Experimental evidence of chiral symmetry breaking in Kekulé-ordered graphene. Phys Rev Lett 126(20):206804. https://doi.org/10.1103/PhysRevLett.126.206804
14. Nomura K, Ryu S, Lee DH (2009) Field-induced Kosterlitz-Thouless transition in the $N = 0$ landau level of graphene. Phys Rev Lett 103:216801
15. Hou CY, Chamon C, Mudry C (2010) Deconfined fractional electric charges in graphene at high magnetic fields. Phys Rev B 81:075427
16. Li SY, Zhang Y, Yin LJ et al (2019) Scanning tunneling microscope study of quantum Hall isospin ferromagnetic states in the zero Landau level in a graphene monolayer. Phys Rev B 100:085437
17. Gutiérrez C, Kim CJ, Brown L et al (2016) Imaging chiral symmetry breaking from Kekulé bond order in graphene. Nat Phys 12:950–958
18. Gamayun OV, Ostroukh VP, Gnezdilov NV et al (2018) Valley-momentum locking in a graphene superlattice with Y-shaped Kekulé bond texture. New J Phys 20:023016
19. Ma D, Fu Z, Sui X et al (2018) Modulating the electronic properties of graphene by self-organized sulfur identical nanoclusters and atomic superlattices confined at an interface. ACS Nano 12(11):10984–10991
20. Gomes KK, Mar W, Ko W et al (2012) Designer Dirac fermions and topological phases in molecular graphene. Nature 483:306–310
21. Giovannetti G, Capone M, van den Brink J et al (2015) Kekulé textures, pseudospin-one Dirac cones, and quadratic band crossings in a graphene-hexagonal indium chalcogenide bilayer. Phys Rev B 91(12):121417
22. Lin Z, Qin W, Zeng J et al (2017) Competing gap opening mechanisms of monolayer graphene and graphene nanoribbons on strong topological insulators. Nano Lett 17(7):4013–4018
23. Virojanadara C, Watcharinyanon S, Zakharov AA et al (2010) Epitaxial graphene on $6H$-SiC and Li intercalation. Phys Rev B 82:205402
24. Sugawara K, Kanetani K, Sato T et al (2011) Fabrication of Li-intercalated bilayer graphene. AIP Adv 1(44):022103
25. Kanetani K, Sugawara K, Sato T et al (2012) Ca intercalated bilayer graphene as a thinnest limit of superconducting C_6Ca. Proc Natl Acad Sci USA 109(48):19610–19613
26. Shimizu R, Sugawara K, Kanetani K et al (2015) Charge-density wave in Ca-intercalated bilayer graphene induced by commensurate lattice matching. Phys Rev Lett 114:146103
27. Caffrey NM, Johansson LI, Xia C et al (2016) Structural and electronic properties of Li-intercalated graphene on SiC(0001). Phys Rev B 93(19):195421
28. Fiori S, Murata Y, Veronese S et al (2017) Li-intercalated graphene on SiC(0001): an STM study. Phys Rev B 96:125429
29. Akiyama R, Takano Y, Endo Y et al (2017) Berry phase shift from 2π to π in bilayer graphene by Li-intercalation and sequential desorption. Appl Phys Lett 110:233106
30. Weller TE, Ellerby M, Saxena SS et al (2005) Superconductivity in the intercalated graphite compounds C_6Yb and C_6Ca. Nat Phys 1(1):39–41

31. Profeta G, Calandra M, Mauri F (2012) Phonon-mediated superconductivity in graphene by lithium deposition. Nat Phys 8:131–134
32. Wang Q, Zhang W, Wang L et al (2013) Large-scale uniform bilayer graphene prepared by vacuum graphitization of 6H-SiC(0001) substrates. J Phys Condens Matter 25(9):095002
33. Varchon F, Feng R, Hass J et al (2007) Electronic structure of epitaxial graphene layers on SiC: effect of the substrate. Phys Rev Lett 99(12):126805
34. Virojanadara C, Zakharov AA, Watcharinyanon S et al (2010) A low-energy electron microscopy and x-ray photo-emission electron microscopy study of Li intercalated into graphene on SiC(0001). New J Phys 12(12):125015
35. Ludbrook BM, Levy G, Nigge P et al (2015) Evidence for superconductivity in Li-decorated monolayer graphene. Proc Natl Acad Sci USA 112(38):11795–11799
36. Pisarski RD (1984) Chiral-symmetry breaking in three-dimensional electrodynamics. Phys Rev D 29:2423–2426
37. Mucha-Kruczyński M, Tsyplyatyev O, Grishin A et al (2008) Characterization of graphene through anisotropy of constant-energy maps in angle-resolved photoemission. Phys Rev B 77:195403
38. Hwang C, Park CH, Siegel DA et al (2011) Direct measurement of quantum phases in graphene via photoemission spectroscopy. Phys Rev B 84:125422
39. Liu Y, Bian G, Miller T et al (2011) Visualizing electronic chirality and Berry phases in graphene systems using photoemission with circularly polarized light. Phys Rev Lett 107:166803
40. Schüler M, De Giovannini U, Hübener H et al (2020) Local Berry curvature signatures in dichroic angle-resolved photoelectron spectroscopy from two-dimensional materials. Sci Adv 6(9): eaay2730

Chapter 5
Coexistence of Flat Band and Kekulé Order

5.1 Research Motivation

In solids, a flat band with a high density of states (DOS) near the Fermi energy E_F can enhance many-body interactions such as electron-phonon (el-ph) interaction and electron-electron correlation, thereby increasing instabilities toward exotic ordered states. For example, in magic-angle twisted bilayer graphene (MABLG) [1], the flat band can be induced through hybridization of the Dirac cones via the moiré superlattice potential and novel phenomena such as superconductivity [2], Mott insulator [3] and charge order [4] have been discovered. Instead of band hybridization near the Dirac point, a different strategy for inducing a flat band is doping graphene to near 1/4 filling, so that the extended flat band near the van Hove singularity (VHS, see Fig. 5.1a for schematic band structure) is shifted to E_F. Graphene near 1/4 filling has attracted extensive interests with intriguing predictions such as instabilities toward *d*-wave superconductivity by repulsive electron-electron interaction [5–7], topological superconductivity [8], and chiral spin density wave [9, 10].

In graphene, the π^* band VHS is at ~ 3 eV above the Dirac point, and an extremely high electron doping is required to shift it to E_F. Experimentally, electron doping of graphene has been achieved by adsorption of alkaline metals [12–17] or Gd, Yb [18–21]. In particular, combining Yb intercalation with K adsorption dopes graphene beyond the VHS [21], however, the Dirac cone of graphene is severely distorted due to the hybridization with the 4f orbital of Yb. We found that the Li-intercalated Kekulé graphene is also a good system for studying the flat band near 1/4 doping in addition to the chiral symmetry breaking, so we use ARPES to study the flat band in the Li-intercalated Kekulé graphene.

© Tsinghua University Press 2023

C. Bao, *Electronic Band Structure Engineering and Ultrafast Dynamics of Dirac Semimetals*, Springer Theses, https://doi.org/10.1007/978-981-99-5325-7_5

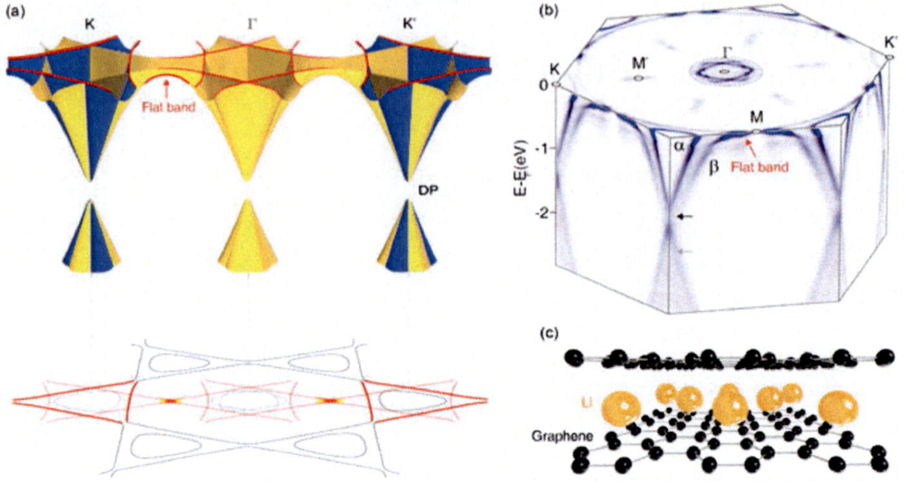

Fig. 5.1 Extended flat band in Li-intercalated bilayer graphene. **a** A schematic for the flat band near the van Hove singularity point in Kekulé-ordered graphene and constant energy maps of the Dirac cones. **b** The full three-dimensional electronic structure of Kekulé-ordered graphene measured by ARPES. The intensity inside the dashed circle around Γ is enhanced by a multiplication factor of 15 for better visualization. **c** A schematic drawing of Li-intercalated bilayer graphene. Reproduced, with permission from [11]

5.2 Extremely High Doping Level

The Li-intercalated bilayer graphene is obtained by intercalating Li to monolayer graphene on SiC substrate [22]. Figure 5.1b shows an overview of the three-dimensional electronic structure from ARPES measurements. Two large pockets labeled as α and β are observed near each Brillouin zone (BZ) corner with the Dirac points at energies of -1.27 and -1.82 eV, indicating a higher doping level than previously reported Li-intercalated graphene [16, 22–24]. The Dirac point energy of -1.82 eV is at an even lower energy than the -1.6 eV in Yb-intercalated graphene [21]. The absence of hybridization is in overall agreement with the calculated electronic structure [25] and distinguishes it from Ca-intercalated graphene [14, 15, 25] or Yb-intercalated graphene [21]. Such high electron doping is made possible by monitoring the electronic structure with ARPES measurements in real-time during Li intercalation, which allows for achieving the optimum doping level.

Figure 5.2a shows the Fermi surface map, from which the carrier concentrations can be directly extracted from the size of the Fermi pockets through the Luttinger theorem [26]. The extracted carrier concentrations are 1.3×10^{14} cm^{-2} and 4.3×10^{14} cm^{-2} for α and β pockets respectively, which are equivalent to 0.07 and 0.23 electrons per unit cell. The high carrier concentration is in line with the superdense Li intercalation between two graphene sheets with an estimated carrier concentration of 4×10^{14} cm^{-2} [27], and here we can resolve each electron pocket and extract the corresponding carrier concentration directly.

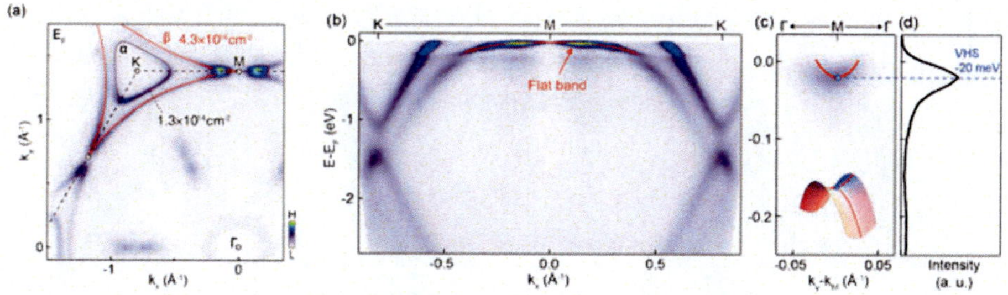

Fig. 5.2 Experimental observation of Flat band. **a** Fermi surface map. **b** Dispersion image along the K-M-K direction. **c** Dispersion image along the Γ-M-Γ direction. The inset is a schematic of the saddle point. **d** EDC at the M point, showing that the VHS (blue dot) is at −20 meV. Reproduced, with permission from [11]

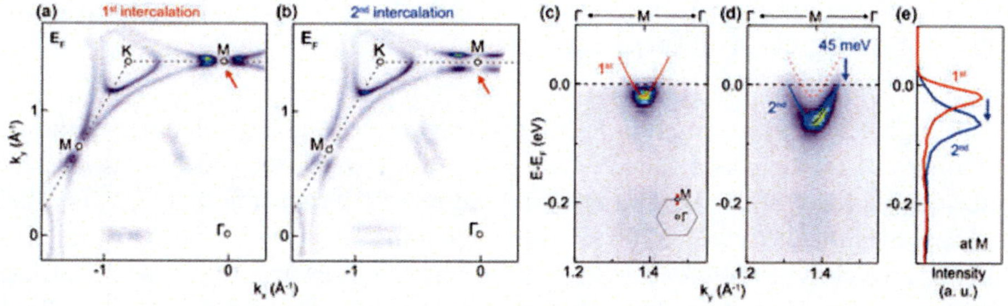

Fig. 5.3 Higher doping level by second intercalation. **a**, **b** Fermi surface maps after first and second intercalations. **c**, **d** Dispersion images at M points after first and second intercalation. **e** Comparison of EDCs at the M point. Reproduced, with permission from [11]

Figure 5.2b, c shows dispersion images measured along two high symmetry directions through the M point. The dispersion is quite flat when approaching the M point and extended in a large momentum range. This corresponds to the top of the valence band along the K-M direction (Fig. 5.2b) and the bottom of the conduction band along the Γ-M direction (Fig. 5.2c), indicating that it is a saddle point (schematic in the inset of Fig. 5.2c). The energy distribution curve (EDC) at the M point shows that the saddle point VHS is at −20 meV, thus Li-intercalated graphene provides an important platform for investigating flat band induced instabilities around the 1/4 filling [8–10]. Importantly, the doping beyond the VHS and the absence of hybridization or orbital contribution from the intercalated Li, allow probing the many-body interactions of the intrinsic Dirac fermions.

Even higher doping level can be obtained by annealing the intercalated sample at 400 °C to remove the intercalated lithium atoms and repeating the intercalation process again. The second intercalation leads to VHS at −65 meV, which is shifted down by 45 meV (Fig. 5.3d, e) and VHS at the M point becomes more extended with a different Fermi surface pocket (Fig. 5.3b). The carrier concentrations extracted from the size of the pockets are $n'_\alpha = 1.4\times10^{14}$ cm^{-2} and $n'_\beta = 5.1\times10^{14}$ cm^{-2} for the second intercalation.

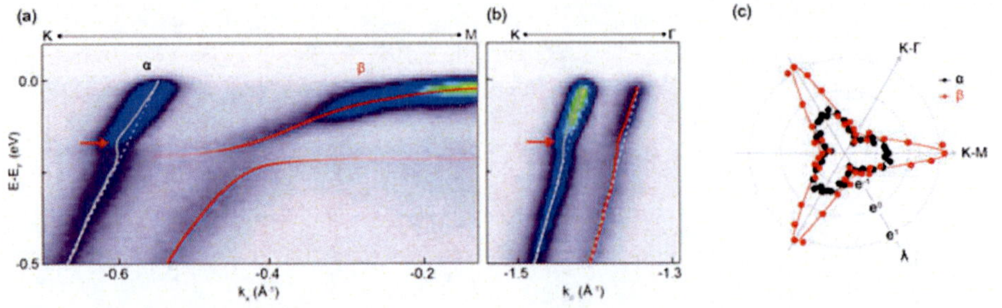

Fig. 5.4 Strong el-ph coupling. **a** Dispersion image around E_F along the K-M direction. **b** Dispersion image around E_F along the K-Γ direction. **c** Extracted angular dependence of el-ph coupling constant of two bands. Reproduced, with permission from [11]

5.3 Strong el-ph Coupling

The high carrier concentration leads to strong el-ph coupling. Figure 5.4a shows a zoom-in of the dispersion image near E_F. For the α pocket which does not reach the doping level of the VHS, a renormalization of the dispersion, a kink, is clearly observed at the energy of -180 ± 20 meV (pointed by the red arrow), while for the β band which is beyond the VHS, the el-ph coupling is even stronger, with a strongly deviated dispersion at similar energy. A comparison of this phonon energy with the calculated phonon dispersion for the Kekulé-ordered graphene [28] shows that the phonons that are coupled with electrons are the in-plane optical phonon A_{1g}.

The el-ph coupling constant can be extracted from the velocity renormalization: $\lambda = v_b/v_F - 1$, where v_b is the bare band velocity and v_F is the renormalized Fermi velocity [29]. v_b and v_f are extracted from the slope of linear fitting for the dispersion below (−0.3–−0.5 eV) and above (−0.13–0 eV) the kink energy. The extracted el-ph coupling strength from the renormalization of the Fermi velocity is $\lambda_\alpha^{KM} = 0.55$ along the K-M direction for the α pocket, and $\lambda_\beta^{KM} = 3.19$ for the β pocket. The el-ph coupling constant is quite anisotropic, and the el-ph coupling strength is reduced to $\lambda_\alpha^{K\Gamma} = 0.25$ and $\lambda_\beta^{K\Gamma} = 0.27$ along the K-Γ direction (Fig. 5.4b). The anisotropic ratio increases from 2.2 for the α pocket (black symbols in Fig. 5.4c) to 11.8 for the β pocket (red symbols in Fig. 5.4c) (see Fig. 5.5 for more data).

The el-ph coupling constant is larger than previously reported values of $\lambda = 0.14$ to 0.61 for pristine graphene [29], alkaline metal doped graphene [30, 31], superconducting Li or Ca doped graphene [25, 32] and graphite [33], due to the higher carrier concentration here. Such strong el-ph coupling may favor instabilities such as superconductivity in graphene and graphite [33–35]. For example, superconductivity with T_c of up to 8.1 K has been predicated in monolayer LiC_6 due to the enhanced el-ph coupling [25]. Experimentally, the superconducting gap has been reported in Li-doped graphene [32] and the Meissner effect has been reported in Li-intercalated graphene [36]. Here, the observation of stronger el-ph coupling in Li-intercalated graphene provides new possibilities for achieving superconductivity with a higher T_c, and future experimental efforts such as transport measurements are needed.

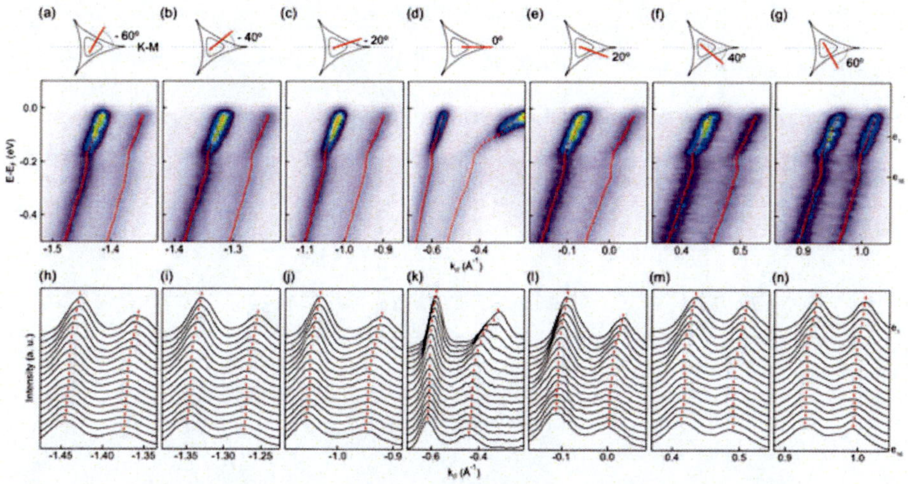

Fig. 5.5 Anisotropic el-ph coupling. **a–g** Dispersion images along different directions as indicated by the red line in schematics. Black curves in schematics are Fermi surface pockets of Dirac cone. Red dots are extracted dispersions from MDCs. **h–n** MDCs for data in (**a**)–(**g**). Reproduced, with permission from [11]

5.4 Co-development of Flat Band and Kekulé Order

To further investigate if there is any correlation between the extended flat band and Kekulé order, we show in Fig. 5.6 the evolution of the electronic structures during the Li intercalation process. Here the size of the Fermi pocket (Fig. 5.6a–f) is used to quantify the carrier concentration, while the dispersions near K (middle panels) and near Γ (Fig. 5.6g–l) are used to monitor the extended flat band and the Kekulé order respectively. Upon Li deposition, the carrier concentration increases, and at carrier concentration of $n = 0.9 \times 10^{14}$ cm^{-2} (see Fig. 5.6c), a new Dirac cone appears, indicating decoupling of the buffer layer from the SiC substrate and the formation of two graphene layers [24]. More Li deposition leads to increasing electron doping, and the replica Dirac cone at the Γ point becomes detectable at a carrier concentration of $n = 3.2 \times 10^{14}$ cm^{-2} (Fig. 5.6d), indicating the formation of the Kekulé order. Further Li deposition leads to a larger β pocket (Fig. 5.6e), and the flat band intensity becomes stronger and finally saturated at $n = 5.6 \times 10^{14}$ cm^{-2} (Fig. 5.6f). The emergence of both the extended flat band and replica Dirac cones near Γ suggests that they not only coexist, but also are strongly connected.

To further confirm the co-development of the extended flat band and the Kekulé order, Fig. 5.7a shows the continuous evolution of the electronic structure during Li intercalation, and quantitative analysis of the intensity for the extended flat band and the replica Dirac cone is shown in Fig. 5.7c. A comparison of the temporal evolution of the intensity for the extended flat band (red triangle in Fig. 5.7c) and replica Dirac cone (blue circle) shows that they have the same trend, and they both increase with the

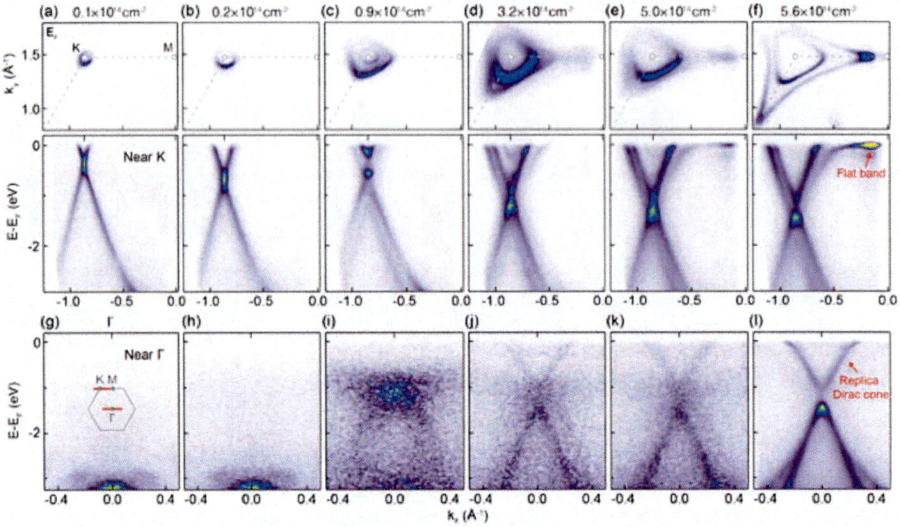

Fig. 5.6 Doping evolution of flat band and Kekulé order. **a–f** Fermi surface maps around the K point and dispersion images along the K-M direction at different doping levels (total carrier concentration). **g–l** Dispersion images along the Γ-K direction at different doping levels the same as (**a–f**). The inset in (**g**) indicates the directions of dispersion images in (**a–f**) and (**g–l**). Reproduced, with permission from [11]

carrier concentration of the β pocket (red symbol in Fig. 5.7b), thereby confirming the coexistence and co-development of the flat band and the Kekulé order.

To understand the coexistence of the Kekulé order and the extended flat band, we first discuss the possible mechanisms of the Kekulé order. There are three main types of mechanisms for the Kekulé order: Friedel oscillation mediated adatom interaction, screened Coulomb interaction between adatoms, and adatom-substrate interaction. The Friedel oscillation mediated adatom interaction is the most well-known mechanism for realizing hidden Kekulé order in graphene with dilute adatoms [37, 38], where the adatoms interact with each other coherently through the Friedel oscillation, and the charge density oscillates with a wavelength $\lambda = \pi / k_F$ (k_F is the Fermi wave vector) to form ordered states. For pristine graphene, k_F is equal to Γ-K distance and λ is equal to the period of Kekulé order, thereby favoring the formation of the Kekulé order. However, this picture becomes invalid at high doping, because k_F strongly deviates from Γ-K distance and λ is no longer equal to the period of the Kekulé order, making it an unlikely mechanism for our Li-intercalated graphene.

The more likely mechanism for the Kekulé order is the direct Coulomb repulsion, complemented by adatom-substrate interaction. The Coulomb repulsion interaction would be screened by graphene, leading to the smaller effective charge and faster decay, however, since the intercalated Li atoms are highly ionized, a stronger Coulomb interaction is expected, increasing the possibility of forming ordered states [39]. This, however, is not sufficient to determine the Kekulé order, and adatom-substrate interaction needs to be taken into account, in order to enforce the adatoms

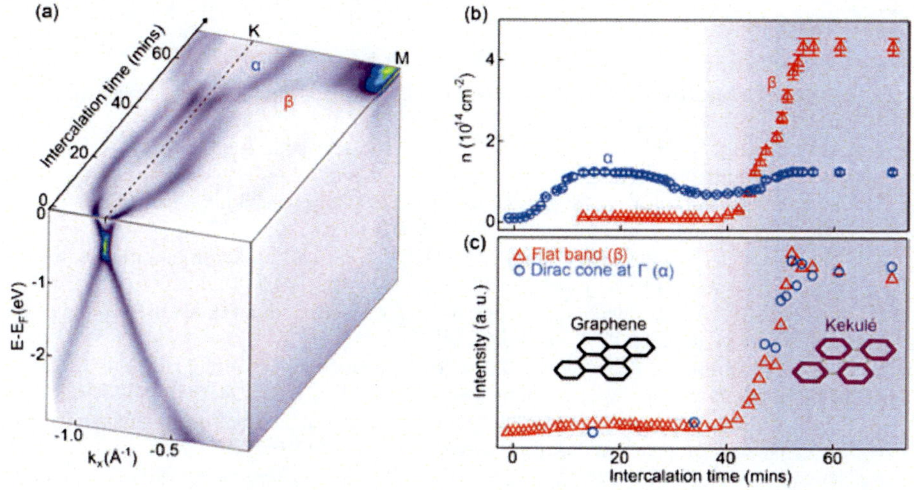

Fig. 5.7 Co-development of flat band and Kekulé order. **a** Continuous evolution of the band structures during Li intercalation. **b** Evolution of carrier concentration for two bands α and β during Li intercalation. **c** Intensity evolution of flat band and replica Dirac cone during Li intercalation. Reproduced, with permission from [11]

to stay in specific positions to maintain the lowest energy, namely, the templating effect [40, 41]. Our first-principles calculations show the hollow site of graphene has the lowest energy (−1.06 eV) for a Li atom than the top (−0.87 eV) or bridge (−0.93 eV) site, and therefore Li atoms prefer the hollow sites and modify the C–C bonding surrounding them, leading to Kekulé-O type bond texture. Therefore, the screened Coulomb interaction and adatom-substrate interaction jointly lead to the formation of the Kekulé order in the Li-intercalated graphene.

Along this line, the co-development of the extended flat band and the Kekulé order can now be understood as the following. The emergence of the flat band relies on the large charge transfer from Li, while the residual positive charge in the Li ions enhances the screened Coulomb repulsion interaction, thus favoring the formation of the Kekulé order. In this way, the extended flat band and the Kekulé order are intrinsically linked by the charge transfer which contributes to the large doping and the induced Coulomb interaction.

5.5 Conclusion

In summary, we report the coexistence of the extended flat band and the Kekulé order in Li-intercalated graphene, thereby identifying an intriguing platform for investigating both electronic instabilities [5–10] and Kekulé order related physics [42, 43]. In particular, their interplay may lead to new complex ordered phases by coupling the nesting vector [9, 10] and the Kekulé order vector and intriguing Kekulé superconductivity [44].

References

1. Bistritzer R, MacDonald AH (2011) Moiré bands in twisted double-layer graphene. Proc Natl Acad Sci USA 108:12233–12237
2. Cao Y, Fatemi V, Fang S et al (2018) Unconventional superconductivity in magic-angle graphene superlattices. Nature 556:43
3. Cao Y, Fatemi V, Demir A et al (2018) Correlated insulator behaviour at half-filling in magic-angle graphene superlattices. Nature 556(7699):80–84
4. Jiang Y, Lai X, Watanabe K et al (2019) Charge order and broken rotational symmetry in magic-angle twisted bilayer graphene. Nature 573(7772):91–95
5. Black-Schaffer AM, Doniach S (2007) Resonating valence bonds and mean-field d-wave superconductivity in graphite. Phys Rev B 75:134512
6. Raghu S, Kivelson SA, Scalapino DJ (2010) Superconductivity in the repulsive Hubbard model: an asymptotically exact weak-coupling solution. Phys Rev B 81:224505
7. Pathak S, Shenoy VB, Baskaran G (2010) Possible high-temperature superconductivity with a $d + id$ pairing symmetry in doped graphene. Phys Rev B 81:085431
8. Nandkishore R, Levitov LS, Chubukov AV (2012) Chiral superconductivity from repulsive interactions in doped graphene. Nat Phys 8:158
9. Wang WS, Xiang YY, Wang QH et al (2012) Functional renormalization group and variational Monte Carlo studies of the electronic instabilities in graphene near $\frac{1}{4}$ doping. Phys Rev B 85:035414
10. Jiang S, Mesaros A, Ran Y (2014) Chiral spin-density wave, spin-charge-chern liquid, and $d + id$ superconductivity in $1/4$-doped correlated electronic systems on the honeycomb lattice. Phys Rev X 4:031040
11. Bao C, Zhang H, Wu X et al (2022) Coexistence of extended flat band and Kekulé order in Li-intercalated graphene. Phys Rev B 105(16):L161106. https://doi.org/10.1103/PhysRevB.105.L161106
12. McChesney JL, Bostwick A, Ohta T et al (2010) Extended van Hove singularity and superconducting instability in doped graphene. Phys Rev Lett 104:136803
13. Sugawara K, Kanetani K, Sato T et al (2011) Fabrication of Li-intercalated bilayer graphene. AIP Adv 1(2):022103
14. Kanetani K, Sugawara K, Sato T et al (2012) Ca intercalated bilayer graphene as a thinnest limit of superconducting C_6Ca. Proc Natl Acad Sci USA 109(48):19610–19613
15. Shimizu R, Sugawara K, Kanetani K et al (2015) Charge-density wave in Ca-intercalated bilayer graphene induced by commensurate lattice matching. Phys Rev Lett 114:146103
16. Caffrey NM, Johansson LI, Xia C et al (2016) Structural and electronic properties of Li-intercalated graphene on SiC(0001). Phys Rev B 93(19):195421
17. Hell MG, Ehlen N, Senkovskiy BV et al (2018) Resonance raman spectrum of doped epitaxial Graphene at the Lifshitz transition. Nano Lett 18(9):6045–6056
18. Link S, Forti S, Stöhr A et al (2019) Introducing strong correlation effects into graphene by gadolinium intercalation. Phys Rev B 100:121407(R)
19. Hwang C, Kim DY, Siegel DA et al (2014) Ytterbium-driven strong enhancement of electron-phonon coupling in graphene. Phys Rev B 90:115417
20. Rosenzweig P, Karakachian H, Link S et al (2019) Tuning the doping level of graphene in the vicinity of the van hove singularity via ytterbium intercalation. Phys Rev B 100:035445
21. Rosenzweig P, Karakachian H, Marchenko D et al (2020) Overdoping graphene beyond the van hove singularity. Phys Rev Lett 125:176403
22. Virojanadara C, Watcharinyanon S, Zakharov AA et al (2010) Epitaxial graphene on $6H$-SiC and Li intercalation. Phys Rev B 82:205402
23. Sugawara K, Kanetani K, Sato T et al (2011) Fabrication of Li-intercalated bilayer graphene. AIP Adv 1(44):022103
24. Bisti F, Profeta G, Vita H et al (2015) Electronic and geometric structure of graphene/SiC(0001) decoupled by lithium intercalation. Phys Rev B 91:245411

25. Profeta G, Calandra M, Mauri F (2012) Phonon-mediated superconductivity in graphene by lithium deposition. Nat Phys 8:131–134
26. Luttinger JM (1960) Fermi surface and some simple equilibrium properties of a system of interacting Fermions. Phys Rev 119:1153–1163
27. Kühne M, Börrnert F, Fecher S et al (2018) Reversible superdense ordering of lithium between two graphene sheets. Nature 564(7735):234–239
28. Zhang H, Bao C, Schuler M et al (2021) Self-energy dynamics and mode-specific phonon threshold effect in a Kekulé-ordered graphene. Natl Sci Rev 9(5):nwab175
29. Zhou SY, Siegel DA, Fedorov AV et al (2008) Kohn anomaly and interplay of electron-electron and electron-phonon interactions in epitaxial graphene. Phys Rev B 78(19):193404
30. Bostwick A, Ohta T, Seyller T et al (2007) Quasiparticle dynamics in graphene. Nat Phys 3(1):36–40
31. Fedorov AV, Verbitskiy NI, Haberer D et al (2014) Observation of a universal donor-dependent vibrational mode in graphene. Nat Commun 5:3257
32. Ludbrook BM, Levy G, Nigge P et al (2015) Evidence for superconductivity in Li-decorated monolayer graphene. Proc Natl Acad Sci USA 112(38):11795–11799
33. Valla T, Camacho J, Pan ZH et al (2009) Anisotropic electron-phonon coupling and dynamical nesting on the graphene sheets in superconducting CaC_6 using angle-resolved photoemission spectroscopy. Phys Rev Lett 102:107007
34. Pan ZH, Camacho J, Upton M et al (2011) Electronic structure of superconducting KC_8 and nonsuperconducting LiC_6 graphite intercalation compounds: evidence for a graphene-sheet-driven superconducting state. Phys Rev Lett 106:187002
35. Yang SL, Sobota JA, Howard CA et al (2014) Superconducting graphene sheets in CaC_6 enabled by phonon-mediated interband interactions. Nat Commun 5:3493
36. Tiwari AP, Shin S, Hwang E et al (2017) Superconductivity at 7.4 K in few layer graphene by Li-intercalation. J Phys Condens Matter 29(44):445701
37. Cheianov VV, Fal'ko VI, Syljuåsen O et al (2009) Hidden Kekulé ordering of adatoms on graphene. Solid State Commun 149(37):1499–1501
38. Gutiérrez C, Kim CJ, Brown L et al (2016) Imaging chiral symmetry breaking from Kekulé bond order in graphene. Nat Phys 12:950–958
39. Song CL, Sun B, Wang YL et al (2012) Charge-transfer-induced cesium superlattices on graphene. Phys Rev Lett 108:156803
40. Li JL, Jia JF, Liang XJ et al (2002) Spontaneous assembly of perfectly ordered identical-size nanocluster arrays. Phys Rev Lett 88:066101
41. Ma D, Fu Z, Sui X et al (2018) Modulating the electronic properties of graphene by self-organized sulfur identical nanoclusters and atomic superlattices confined at an interface. ACS Nano 12(11):10984–10991
42. Hou CY, Chamon C, Mudry C (2007) Electron fractionalization in two-dimensional graphene-like structures. Phys Rev Lett 98:186809
43. Andrade E, Carrillo-Bastos R, Naumis GG (2019) Valley engineering by strain in Kekulé-distorted graphene. Phys Rev B 99:035411
44. Roy B, Herbut IF (2010) Unconventional superconductivity on honeycomb lattice: theory of Kekulé order parameter. Phys Rev B 82:035429

Chapter 6
Dirac Fermion Cooling in 3D Dirac Semimetal Cd_3As_2

6.1 Research Motivation

Three-dimensional (3D) Dirac semimetals [1, 2] are characterized by Dirac fermions with linear dispersion near the point-like Dirac nodes in the 3D momentum space, which can be viewed as 3D analogues of two-dimensional (2D) Dirac fermions in graphene [3, 4]. Such unique electronic structure strongly affects the light-matter interaction, and can lead to intriguing optoelectronic properties with potential applications in devices such as high-performance photodetectors [5]. Moreover, the vanishing density of states (DOS) near the Dirac point suppresses the scattering phase space, and the slower interband relaxation could lead to population inversion, which is highly desirable for broad-bandwidth Terahertz or mid-infrared lasing application [6] and realizing exotic dynamical excitonic insulator [7]. So far, Dirac fermion cooling dynamics has been widely investigated for 2D graphene, and a population inversion with a timescale of ∼130 fs has been reported [8]. However, how the cooling of 3D Dirac fermions differs from 2D Dirac fermions and whether it is possible to host a long-lived population inversion are fundamental questions that are still awaiting to be answered.

As a model 3D Dirac semimetal, Cd_3As_2 has attracted extensive research interests [9–13] with a large Fermi velocity, high carrier mobility, good chemical stability, and it shows broadband responses for ultrafast nano-optoelectronics [14, 15]. The understanding of the fundamental ultrafast carrier dynamics upon photoexcitation is therefore important for its nano-optoelectronics device applications. So far, the ultrafast dynamics of 3D Dirac fermions have been investigated by time-resolved optical measurements [14–22], and two main relaxation processes involving a faster one in 0.4–4 ps [16, 17] and a slower one from 2 to 8 ps [16, 17, 20] have been reported. Recently, a possible long-lived population inversion has been suggested from the ultrafast response of THz photoconductivity [21]. We note that the optical response is contributed by all photocarriers at different electron energy and momentum, making it difficult to identify the contribution of electrons from different parts of the Dirac cone. Resolving the Dirac fermion dynamics with both energy- and

© Tsinghua University Press 2023
C. Bao, *Electronic Band Structure Engineering and Ultrafast Dynamics of Dirac Semimetals*, Springer Theses, https://doi.org/10.1007/978-981-99-5325-7_6

momentum-resolved information will allow extracting the chemical potential, electronic temperature, energy-resolved relaxation rate, etc., which are critical for understanding the Dirac fermion cooling dynamics and confirming whether there is population inversion. In addition, photo-induced topological phase transitions have been proposed for 3D topological semimetals [6, 23–26], and accessing the dynamics of such 3D topological semimetals would be the first important step.

Time- and angle-resolved photoemission spectroscopy (TrARPES) is a powerful technique for revealing the ultrafast dynamics of Dirac fermions with both energy- and momentum-resolved information [27, 28]. While this technique has been applied to various quasi-2D materials such as high-temperature superconductors [29–31], topological insulators [32, 33] and graphene [8, 34] to reveal rich dynamics information, TrARPES study of 3D Dirac fermions has been missing so far due to the lack of widely tunable probe photon energy, which is required to scan through the conical dispersion at the desired k_z precisely. Here by using TrARPES with a tunable probe photon source from 5.3 to 7.0 eV, we are able to reveal the transient dynamics of 3D Dirac fermions in Cd_3As_2 upon photoexcitation.

6.2 Revealing the 3D Dirac Cone

The 3D Dirac cones emerge at two isolated Dirac nodes at $k_z = \pm 0.065$ c* in the k_x-k_y-k_z momentum space [10–12]. In order to access the Dirac node, a tunable probe photon energy covering the right k_z value is required. A calculation using an inner potential of 10.6 eV [10] shows that k_z of the Dirac node (0.065 c*) corresponds to a probe photon energy of 6.7 eV. A TrARPES system with a tunable probe photon energy covering 6.7 eV is therefore required for revealing the dynamics of the Dirac electrons in Cd_3As_2.

In TrARPES, the maximum photon energy generated from the fourth harmonic generation (FHG) of the BBO crystal is 6.3 eV [29, 35], which is not sufficient to access the Dirac node yet. KBBF crystal can extend the FHG output to 7.0 eV [36], and a continuously tunable photon source covering 5.6–7.0 eV has been used in static laser-based ARPES measurements [37, 38]. Here, the developed time-resolved version facilitates ultrafast dynamic studies of 3D materials, allowing to probe the carrier dynamics near the 3D Dirac node.

The Cd_3As_2 single crystal is cleaved at (111) surface as shown in Fig. 6.1 which is confirmed by Laue (Fig. 6.1a) and LEED (Fig. 6.1b) patterns. The ARPES measurement is performed along Γ-M as indicated in Fig. 6.1c. Figure 6.2b shows ARPES dispersion images measured at photon energies from 6.93 to 5.75 eV at 80 K. The conical-shaped dispersion is clearly observed at 6.70 eV, whose intensity decreases when detuned from this photon energy and becomes undetectable at 5.75 eV, confirming its 3D nature. The Dirac cone appears at the photon energy of 6.70 eV with the corresponding $k_z = 0.065$ c*, consistent with previous ARPES studies using synchrotron light sources [10–12].

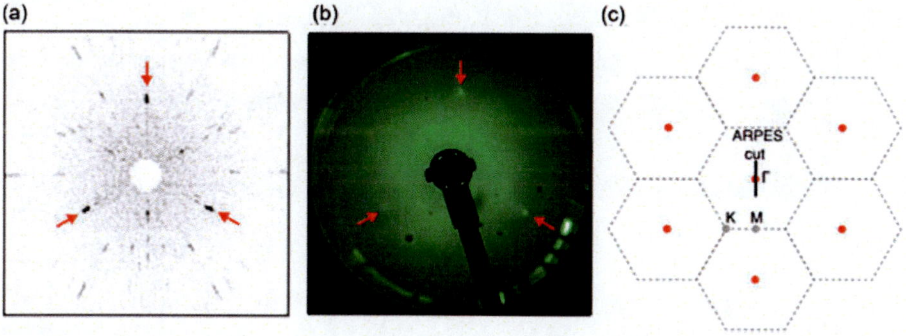

Fig. 6.1 Crystal orientation. **a** Laue pattern of (111) surface. **b** LEED pattern of (111) surface. **c** Projected surface Brillouin zone of (111) surface and schematics for ARPES cut direction. Reproduced, with permission from [39]

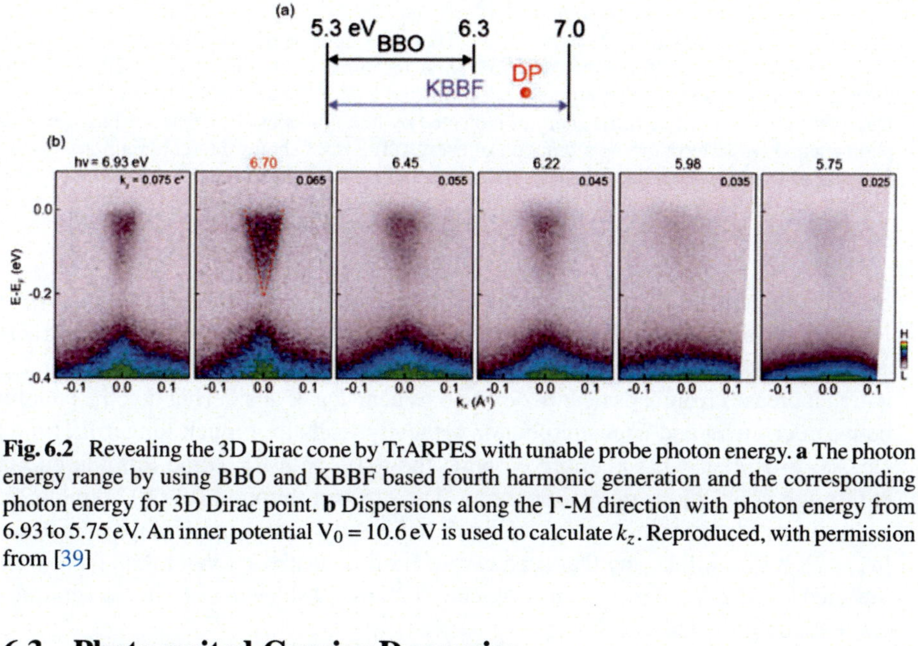

Fig. 6.2 Revealing the 3D Dirac cone by TrARPES with tunable probe photon energy. **a** The photon energy range by using BBO and KBBF based fourth harmonic generation and the corresponding photon energy for 3D Dirac point. **b** Dispersions along the Γ-M direction with photon energy from 6.93 to 5.75 eV. An inner potential $V_0 = 10.6$ eV is used to calculate k_z. Reproduced, with permission from [39]

6.3 Photoexcited Carrier Dynamics

Using the selected probe photon energy of 6.70 eV, the dynamics of 3D Dirac fermions is revealed by TrARPES measurements. Figure 6.3a–e show TrARPES snapshots of the Dirac cone at different delay times. Photoexcitation of electrons into the unoccupied states up to 0.2 eV above E_F (indicated by red color in the differential images in Fig. 6.3g–j) and photoexcited holes below E_F (indicated by blue color) are observed as schematically illustrated in Fig. 6.3f. The photoexcited electrons above 0.1 eV relax much faster and recover at 5 ps (Fig. 6.3d), while those near E_F return to the equilibrium state at a later delay time and completely recover by 25 ps (Fig. 6.3e). The lifetimes of photoexcited carriers at different energies can

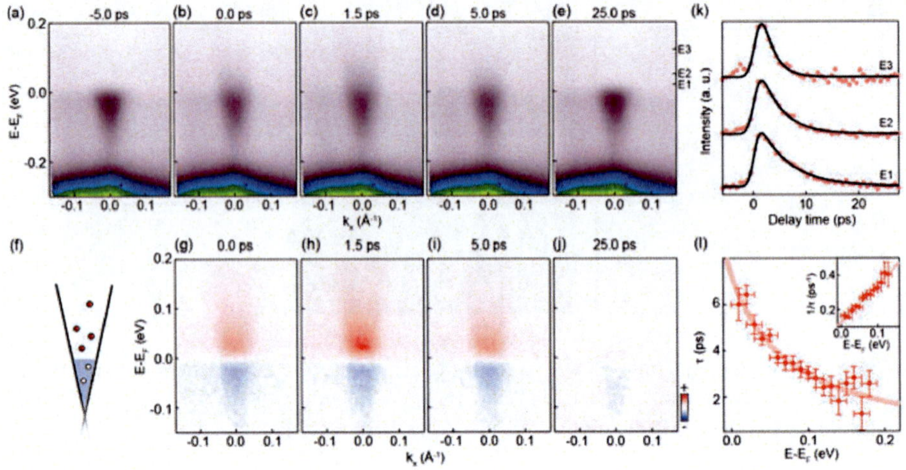

Fig. 6.3 Photoexcited carrier dynamics of 3D Dirac fermions. **a–e** TrARPES spectra at different delay times. **f** Schematics for photoexcited 3D Dirac fermions. **g–j** Differential TrARPES spectra at different delay times after subtracting the spectrum at −5 ps. **k** Temporal evolution of photoexcited Dirac fermions at a few selected energies indicated by tick marks in (**e**). **l** Extracted lifetimes τ of photoexcited Dirac fermions as a function of energy. The inset shows the scattering rate $1/\tau$ as a function of energy with a linear fitting. The pump and probe photon energies are 6.7 and 1.68 eV respectively, with a pump fluence of 40 uJ/cm^2. Reproduced, with permission from [39]

be extracted by fitting the temporal evolution of TrARPES spectra as shown in Fig. 6.3k. A single-exponential function is sufficient to fit the data. The extracted energy-dependent lifetime of photocarriers excited above E_F is plotted in Fig. 6.3l, which increases from 2 ps near 0.2 eV to 6 ps near E_F. Considering that the coupling between electron and acoustic phonon typically leads to a much longer lifetime in Cd_3As_2 [40] than what is observed here, the main path for Dirac fermion cooling is through coupling to optical phonons. The electron-phonon (el-ph) coupling has been shown to play an important role in the carrier dynamics in topological materials [41, 42]. It is worth noting that such carrier lifetime is much longer than the lifetime reported in 2D Dirac fermion in graphene (0.25 ps) [43] because of the lower optical phonon energy in Cd_3As_2 [44].

To extract the el-ph coupling strength g, we replot the scattering rate (inversely proportional to the relaxation lifetime) as a function of energy in the inset of Fig. 6.3l. A linear energy dependence of scattering rate is clearly identified, and the electron-phonon coupling strength g can be extracted from the slope, which equals to $\frac{g(T_e-T_L)}{2C_e k_B T_e}$ [18, 45], where C_e is the electron heat capacity, T_e is the transient electronic temperature and T_L is the lattice temperature. The slope of 2.1 ± 0.1 ps^{-1}·eV^{-1} gives an el-ph coupling strength $g = (6.2 \pm 0.3) \times 10^{15}$ W·K^{-1}m^{-3}, where $C_e = 70$ J·K^{-2}m$^{-3} \cdot T_e$ [18], $T_e = 450$ K (extracted from the Fermi-Dirac distribution) and T_L = 80 K (sample temperature). We note that similar wavelength-dependent relaxation rates with a more complex relation deviating from linearity have been observed by time-resolved optical measurements [15, 18, 22] where different electronic states in the Dirac cone are selectively excited by tuning the photon energy. Here, the direct energy- and momentum-resolving capability allows to access each electronic state in the 3D Dirac cone directly and to extract the el-ph coupling strength more accurately.

6.4 Ultrafast Thermodynamics

Such direct snapshots of photoexcited 3D Dirac fermions allow revealing not only the photo-carrier dynamics but also the thermodynamic evolution [46–48], which could provide further insight into the underlying physics, in particular, whether there is population inversion. Figure 6.4a shows the evolution of photoelectrons as a function of energy and delay time. Thermodynamic parameters including the transient electronic temperature T_e and chemical potential μ can be extracted by fitting the integrated EDCs (Fig. 6.4a) with a Fermi-Dirac distribution as illustrated in Fig. 6.4b, where the edge position defines μ while the width of the edge is determined by $4kT_e$. Figure 6.4c shows EDCs at a few selected delay times, and the extracted T_e and μ are plotted in Fig. 6.4d, e. The cooling of the electronic temperature is through the energy transfer from hot electrons to the cold lattice, which could be described by the two-temperature model (TTM) [49]

$$C_e \frac{dT_e}{dt} = -g(T_e - T_L), \tag{6.1}$$

the integration of which gives

$$T_e(t) = (T_e(0) - T_L)e^{-\frac{t}{C_e/g}} + T_L \tag{6.2}$$

where $T_e(0)$ is electronic temperature at $t = 0$. Analysis of the electronic temperature shows a single-exponential relaxation consistent with TTM and gives a relaxation time τ_T of 4.8±0.3 ps, providing the important characteristic time of el-ph scattering in Cd_3As_2. In addition, the el-ph coupling strength g can also be extracted from the relaxation time τ_T by $g = C_e/\tau_T = (6.6 \pm 0.4) \times 10^{15}$ W$\cdot$ K^{-1}m^{-3}, which is consistent with the value extracted from photo-carrier dynamics discussed above, thus confirming the validity of the thermodynamic analysis.

6.5 Population Inversion

The chemical potential also shows a shift with delay times, which can be used to check if there is any population inversion when combined with the electronic temperature and density of states analysis. The increase of the chemical potential is identified from the shift of the edge in Fig. 6.4c and the extracted values with a maximum value of 6 meV are plotted in Fig. 6.4e. To check if the increased chemical potential

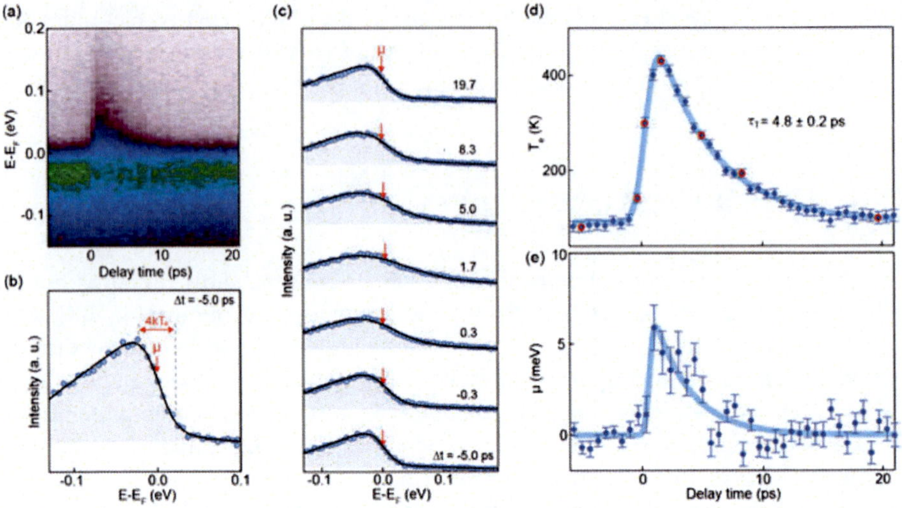

Fig. 6.4 Ultrafast thermodynamic evolution of 3D Dirac fermions. **a** Integrated EDCs as a function delay time. **b** EDC at delay time of -5.0 ps and corresponding Fermi-Dirac fitting. **c** Representative EDCs at delay times of −5.0, −0.3, 0.3, 1.7, 5.0, 8.3 and 19.7 ps and black curves are Fermi-Dirac distribution fittings. **d** Extracted electronic temperature as a function of delay time. The data points extracted from the curves in panel (**c**) are highlighted by red color. **e** Extracted chemical potential as a function of delay time. Reproduced, with permission from [39]

implies the population inversion, three possible mechanisms are examined, including electron-hole asymmetry, heating effects, and carrier concentration change. First, electron-hole asymmetry [50] can be excluded, because the thermal excitation energy ($kT_e = 39$ meV for $T_{e,max} = 450$ K as shown in Fig. 6.4d) is much smaller than the binding energy of the Dirac point (~200 meV), and therefore no electron can be excited thermally from the VB to CB and no hole can be generated in VB. Secondly, the heating effects can also be excluded. The increase of T_e means that more electrons are excited to higher energy, meanwhile, the DOS for the CB increases with the energy for 3D Dirac fermions as DOS(E) $\propto$ E^2. If the number of carriers is conserved for the CB, a lower μ would be expected (see the simulation in Fig. 6.5a), which is in contrast to our experimental results. Therefore, the increase of chemical potential suggests the carrier concentration for the CB of the 3D Dirac cone is not conserved upon photoexcitation, namely, there is a population inversion with photoexcited electrons in the CB.

To confirm the population inversion, we further extract the temporal evolution of the carrier concentration. Density of states of the conduction band in 3D Dirac semimetal is:

$$DOS(E) = \frac{(E - E_{DP})^2}{\pi^2 \hbar^3 v_x v_y v_z}, \quad (6.3)$$

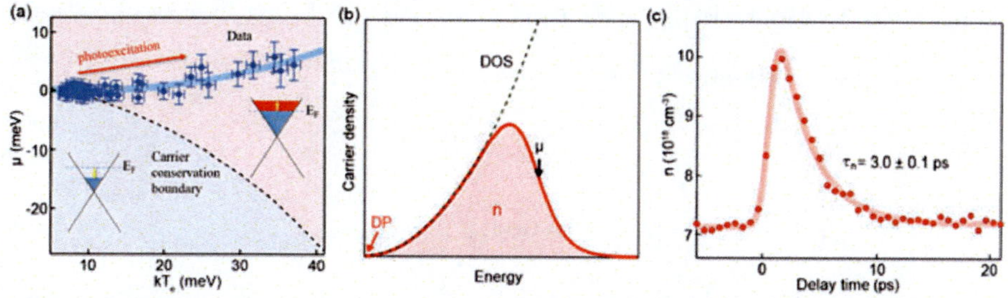

Fig. 6.5 Experimental evidence of population inversion. **a** Replotted μ as a function of kT_e. The black dashed curve is simulated temperature-dependent chemical potential for 3D Dirac fermion under the assumption of carrier conservation. **b** Simulated energy-dependent carrier density of 3D Dirac semimetal to illustrate the calculation of carrier concentration in CB. **c** Carrier concentration in CB as a function of delay time. Reproduced, with permission from [39]

where E_{DP} is the energy of Dirac point and v_x, v_y and v_z are Fermi velocities along k_x, k_y and k_z directions, respectively. Then, the carrier concentration is

$$n = \int_{E_{DP}}^{\infty} \frac{DOS(E)dE}{1 + e^{\frac{E-\mu}{kT_e}}} = \int_{E_{DP}}^{\infty} \frac{(E - E_{DP})^2}{\pi^2 \hbar^3 v_x v_y v_z} \frac{dE}{1 + e^{\frac{E-\mu}{kT_e}}}. \tag{6.4}$$

Therefore, the carrier concentration is obtained from the electronic temperature (T_e) and chemical potential (μ), where $E_{DP} = -0.2$ eV [11, 12], $v_x = 5$ eV·Å, $v_y = 5.1$ eV·Åand $v_z = 1.5$ eV·Å[10] are used.

The temporal evolution of carrier concentration is shown in Fig. 6.5c, which shows a clear increase of carrier concentration in the CB. We note that the change of carrier concentration in the VB, however, cannot be directly extracted from the TrARPES data due to the multi-band nature of the VB [51] and the complications from the other strong band around −0.4 eV, which exists in different k_z values and dominates the intensity of the VB over the intrinsic 3D Dirac cone. Therefore, the change in the carrier concentration of the VB has to be deduced indirectly by considering the change in the VB together with the conservation of total charge carriers. The conservation of total charge carriers is based on the fact that the pump photon energy (1.68 eV) is much smaller than the work function (typically around 4.5 eV), and therefore no electrons can be excited out of the crystal. Therefore, the observation of increased carrier concentration in CB is equivalent to the increase of the same number of holes in VB due to the conservation of carriers, thus confirming the light-induced population inversion. Fitting the relaxation curve in Fig. 6.5c shows that the population inversion in Cd_3 As $_2$ has a lifetime of 3.0±0.1 ps, which is an order of magnitude longer than the population inversion lifetime (130 fs) in graphene [8]. This suggests that the interband relaxation plays an important role in 3D Dirac fermion cooling in addition to el-ph scattering induced intraband relaxation upon photoexcitation as summarized in Fig. 6.6a–c. However, the interband relaxation for 3D Dirac fermions is much slower compared to 2D Dirac fermions [8, 33], where the population inversion exists only in 130 fs [8].

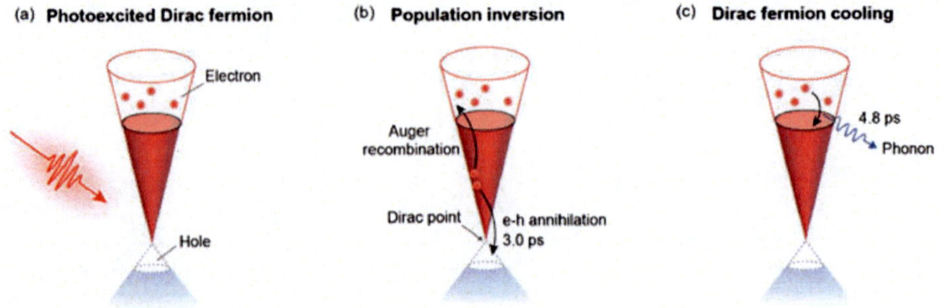

Fig. 6.6 Schematic 3D Dirac fermion relaxation mechanism. **a–c** Schematics for 3D Dirac fermion photoexcitation, population inversion and cooling mechanism. Reproduced, with permission from [39]

The long-lived population inversion in 3D Dirac semimetal is likely related to the suppression of Auger recombination, namely, electron-hole recombination with energy and momentum transfer to another carrier (schematically illustrated in Fig. 6.6b). The Auger recombination probability is determined by both the scattering phase space and the scattering matrix element. For 2D Dirac fermions, although the scattering phase space is vanishing, the scattering matrix element is diverging [52]. This leads to a finite Auger recombination probability, which is a main obstacle for population inversion. For 3D Dirac semimetals, theoretical work suggests that Auger recombination is suppressed because of the vanishing phase space and the finite scattering matrix element [53]. This could lead to an order of magnitude enhancement in the lifetime of population inversion as reported in our TrARPES data.

6.6 Conclusion

In summary, by utilizing the TrARPES system with widely tunable probe photon energy to reveal the photoexcited 3D Dirac fermions directly, the relaxation dynamics of the 3D Dirac fermions and its distinction from 2D Dirac fermions are revealed. Moreover, a population inversion with a lifetime of 3.0 ps is reported. Our work reveals the fundamental dynamics of 3D Dirac fermions and opens up new opportunities for exploring the dynamics of topological semimetals as well as light-induced transient topological phase transitions.

References

1. Armitage NP, Mele EJ, Vishwanath A (2018) Weyl and Dirac semimetals in three-dimensional solids. Rev Mod Phys 90:015001
2. Lv BQ, Qian T, Ding H (2021) Experimental perspective on three-dimensional topological semimetals. Rev Mod Phys 93:025002

3. Geim AK, Novoselov KS (2007) The rise of graphene. Nat Mater 6:183–191
4. Castro Neto AH, Guinea F, Peres NMR et al (2009) The electronic properties of graphene. Rev Mod Phys 81:109–162
5. Liu J, Xia F, Xiao D et al (2020) Semimetals for high-performance photodetection. Nat Mater 19(8):830–837
6. Weber CP (2021) Ultrafast investigation and control of Dirac and Weyl semimetals. J Appl Phys 129(7):070901
7. Pertsova A, Balatsky AV (2020) Dynamically induced excitonic instability in pumped Dirac materials. Ann Phys (Berl) 532(2):1900549
8. Gierz I, Petersen JC, Mitrano M et al (2013) Snapshots of non-equilibrium Dirac carrier distributions in graphene. Nat Mater 12(12):1119–24
9. Wang Z, Weng H, Wu Q et al (2013) Three-dimensional Dirac semimetal and quantum transport in Cd_3As_2. Phys Rev B 88:125427
10. Liu ZK, Jiang J, Zhou B et al (2014) A stable three-dimensional topological Dirac semimetal Cd_3As_2. Nat Mater 13(7):677–81
11. Neupane M, Xu SY, Sankar R et al (2014) Observation of a three-dimensional topological Dirac semimetal phase in high-mobility Cd_3As_2. Nat Commun 5:3786
12. Ali MN, Gibson Q, Jeon S et al (2014) The crystal and electronic structures of Cd_3As_2, the three-dimensional electronic analogue of graphene. Inorg Chem 53(8):4062–4067
13. Liang T, Gibson Q, Ali MN et al (2015) Ultrahigh mobility and giant magnetoresistance in the Dirac semimetal Cd_3As_2. Nat Mater 14(3):280
14. Wang Q, Li CZ, Ge S et al (2017) Ultrafast broadband photodetectors based on three-dimensional Dirac semimetal Cd_3As_2. Nano Lett 17(2):834–841
15. Zhu C, Wang F, Meng Y et al (2017) A robust and tuneable mid-infrared optical switch enabled by bulk Dirac fermions. Nat Commun 8:14111
16. Weber CP, Arushanov E, Berggren BS et al (2015) Transient reflectance of photoexcited Cd_3As_2. Appl Phys Lett 106(23):231904
17. Lu W, Ge S, Liu X et al (2017) Ultrafast relaxation dynamics of photoexcited Dirac fermions in the three-dimensional Dirac semimetal Cd_3As_2. Phys Rev B 95:024303
18. Zhu C, Yuan X, Xiu F et al (2017) Broadband hot-carrier dynamics in three-dimensional Dirac semimetal Cd_3As_2. Appl Phys Lett 111(9):091101
19. Sun F, Wu Q, Wu YL et al (2017) Coherent helix vacancy phonon and its ultrafast dynamics waning in topological Dirac semimetal Cd_3As_2. Phys Rev B 95(23):235108
20. Lu W, Ling J, Xiu F et al (2018) Terahertz probe of photoexcited carrier dynamics in the Dirac semimetal Cd_3As_2. Phys Rev B 98:104310
21. Zhang W, Yang Y, Suo P et al (2019) Ultrafast photocarrier dynamics in a 3D Dirac semimetal Cd_3As_2 film studied with terahertz spectroscopy. Appl Phys Lett 114(22):221102
22. Zhai G, Ma C, Xiang J et al (2020) Mid-infrared transient reflectance study of the Dirac semimetal Cd_3As_2 under strong optical pumping. Phys Rev B 101:174310
23. Ebihara S, Fukushima K, Oka T (2016) Chiral pumping effect induced by rotating electric fields. Phys Rev B 93:155107
24. Chan CK, Oh YT, Han JH et al (2016) Type-II Weyl cone transitions in driven semimetal. Phys Rev B 94:121106
25. Hübener H, Sentef MA, Giovannini UD et al (2017) Creating stable Floquet-Weyl semimetals by laser-driving of 3D Dirac materials. Nat Commun 8:13940
26. Bao C, Tang P, Sun D et al (2022) Light-induced emergent phenomena in 2D materials and topological materials. Nat Rev Phys 4:33–48
27. Smallwood CL, Kaindl RA, Lanzara A (2016) Ultrafast angle-resolved photoemission spectroscopy of quantum materials. Europhys Lett 115(2):27001
28. Sobota JA, He Y, Shen ZX (2021) Angle-resolved photoemission studies of quantum materials. Rev Mod Phys 93:025006
29. Perfetti L, Loukakos P A, Lisowski M et al, Ultrafast electron relaxation in superconducting $Bi_2Sr_2CaCu_2O_{8+\delta}$

30. Graf J, Jozwiak C, Smallwood CL et al (2011) Nodal quasiparticle meltdown in ultrahigh-resolution pump-probe angle-resolved photoemission. Nat Phys 7(10):805–809
31. Rettig L, Cortés R, Thirupathaiah S et al (2012) Ultrafast momentum-dependent response of electrons in antiferromagnetic $EuFe_2As_2$ driven by optical excitation. Phys Rev Lett 108:097002
32. Sobota JA, Yang S, Analytis JG et al (2012) Ultrafast optical excitation of a persistent surface-state population in the topological insulator Bi_2Se_3. Phys Rev Lett 108:117403
33. Wang YH, Hsieh D, Sie EJ et al (2012) Measurement of intrinsic Dirac fermion cooling on the surface of the topological insulator Bi_2Se_3 using time-resolved and angle-resolved photoemission spectroscopy. Phys Rev Lett 109:127401
34. Johannsen JC, Ulstrup S, Cilento F et al (2013) Direct view of hot carrier dynamics in graphene. Phys Rev Lett 111:027403
35. Carpene E, Mancini E, Dallera C et al (2009) A versatile apparatus for time-resolved photoemission spectroscopy via femtosecond pump-probe experiments. Rev Sci Instrum 80(5):055101
36. Zhang H, Wang G, Guo L et al (2008) 175 to 210 nm widely tunable deep-ultraviolet light generation based on KBBF crystal. Appl Phys B 93(2):323–326
37. Jiang R, Mou D, Wu Y et al (2014) Tunable vacuum ultraviolet laser based spectrometer for angle resolved photoemission spectroscopy. Rev Sci Instrum 85(3):033902
38. Zhou X, He S, Liu G et al (2018) New developments in laser-based photoemission spectroscopy and its scientific applications: a key issues review. Rep Prog Phys 81(6):062101
39. Bao C, Li Q, Xu S et al (2022) Population inversion and Dirac fermion cooling in 3D Dirac semimetal Cd_3As_2. Nano Lett 22(3):1138–1144. https://doi.org/10.1021/acs.nanolett.1c04250
40. Lundgren R, Fiete GA (2015) Electronic cooling in Weyl and Dirac semimetals. Phys Rev B 92:125139
41. Osterhoudt GB, Wang Y, Garcia CAC et al (2021) Evidence for dominant phonon-electron scattering in Weyl semimetal WP_2. Phys Rev X 11:011017
42. Clark OJ, Freyse F, Aguilera I et al (2021) Observation of a giant mass enhancement in the ultrafast electron dynamics of a topological semimetal. Commun Phys 4(1):1–11
43. Breusing M, Kuehn S, Winzer T et al (2011) Ultrafast nonequilibrium carrier dynamics in a single graphene layer. Phys Rev B 83:153410
44. Sharafeev A, Gnezdilov V, Sankar R et al (2017) Optical phonon dynamics and electronic fluctuations in the Dirac semimetal Cd_3As_2. Phys Rev B 95:235148
45. Sun BY, Zhou Y, Wu MW (2012) Dynamics of photoexcited carriers in graphene. Phys Rev B 85:125413
46. Ulstrup S, Johannsen JC, Grioni M et al (2014) Extracting the temperature of hot carriers in time-and angle-resolved photoemission. Rev Sci Instrum 85(1):013907
47. Rohde G, Stange A, Müller A et al (2018) Ultrafast formation of a Fermi-Dirac distributed electron gas. Phys Rev Lett 121:256401
48. Shi X, You W, Zhang Y et al (2019) Ultrafast electron calorimetry uncovers a new long-lived metastable state in 1T-$TaSe_2$ mediated by mode-selective electron-phonon coupling. Sci Adv 5(3): eaav4449
49. Allen PB (1987) Theory of thermal relaxation of electrons in metals. Phys Rev Lett 59(13):1460
50. Miller TL, Smallwood CL, Zhang W et al (2015) Photoinduced changes of the chemical potential in superconducting $Bi_2Sr_2CaCu_2O_{8+\delta}$. Phys Rev B 92:144506
51. Roth S, Lee H, Sterzi A et al (2018) Reinvestigating the surface and bulk electronic properties of Cd_3As_2. Phys Rev B 97:165439
52. Winzer T, Knorr A, Malic E (2010) Carrier multiplication in graphene. Nano Lett 10(12):4839–4843
53. Afanasiev AN, Greshnov AA, Svintsov D (2019) Relativistic suppression of Auger recombination in Weyl semimetals. Phys Rev B 99:115202

Chapter 7
Summary and Prospects

This thesis reports the development of a novel TrARPES system with tunable probe photon energy, and an investigation of the band structure engineering of 2D graphene and ultrafast dynamics of a model 3D Dirac semimetal by utilizing this system.

As a 2D Dirac semimetal, the electronic graphene has been widely investigated, and our work advances the electronic structure engineering of graphene by realizing the long-sought-after chiral symmetry breaking by inducing a Kekulé order via Li intercalation. Interestingly, the Kekulé order coexists and codevelops with an extended flat band near E_F with strong electron-phonon interaction, thereby providing a new playground for investigating both CSB related physics and flat band related instabilities, such as superconductivity [1], fractional charge [2] etc. The Kekulé-ordered graphene realized in this thesis shows extremely high carrier density and the flat band near the Fermi energy, so whether it is superconducting would be the next important scientific question to be answered and further transport measurements should be performed. In addition, the Kekulé-ordered graphene is also a rare system hosting both chiral symmetry breaking and flat band, opening up new opportunities for exploring their interplay, for example intriguing Kekulé superconductivity [3] and new ordered state combing Kekulé order and Fermi surface nesting from the flat band [4, 5].

For 3D Dirac/Weyl semimetals which are predicted to exhibit rich interesting physics in the nonequilibrium state, revealing the ultrafast dynamics with energy and momentum resolution has been hindered due to the lack of a suitable TrAREPS system with tunable photon energy to access the 3D Dirac point. By developing a TrARPES system with a highly tunable probe photon energy from 5.3 to 7.0 eV using a BBO + KBBF scheme, we can now access the Dirac point of Cd_3As_2, and reveal the ultrafast dynamics with high energy, momentum and time resolution. A population inversion is reported based on the observation of accumulated photoexcited carriers in the conduction band with a lifetime of 3.0 ps. This work reveals the distinctive relaxation between the relaxation dynamics of 2D graphene and 3D Dirac semimetal and provides direct experimental evidence for a long-lived popu-

© Tsinghua University Press 2023

C. Bao, *Electronic Band Structure Engineering and Ultrafast Dynamics of Dirac Semimetals*, Springer Theses, https://doi.org/10.1007/978-981-99-5325-7_7

lation inversion which is important toward the experimental realization of dynamic excitonic insulator.

We believe that the development of TrARPES with tunable probe photon energy for 3D quantum materials and its demonstration on 3D Dirac semimetal provides new opportunities for exploring light-matter interaction in 3D quantum materials with direct energy and momentum resolutions. For example, the Floquet engineering in 3D topological materials has been predicted to lead to intriguing topological phase transitions [6–14], while experimental progress has been quite limited. Using TrARPES to explore the light-induced topological phase transition would be an important step toward light-induced intriguing states such as light-induced topological phase transition, and light-induced dynamical excitonic insulator transition.

Parts of this Thesis have been Published in the Following Documents:
Journal Publications:

Changhua Bao†, Hongyun Zhang†, Teng Zhang, et al. Experimental evidence of chiral symmetry breaking in a Kekulé-ordered graphene. Phys. Rev. Lett., 2021; 126:206804.

Changhua Bao, Peizhe Tang*, Dong Sun* and Shuyun Zhou* Light-induced emergent phenomena in 2D materials and topological materials. Nat. Rev. Phys., 2022; 4:33.

Changhua Bao†, Hongyun Zhang†, **Xi Wu**, et al. Coexistence of extended flat band and Kekulé order in Li-intercalated graphene. Phys. Rev. B, 2022; 105:L161106.

Changhua Bao, Qian Li, Sheng Xu, et al. Population inversion and Dirac fermion cooling in 3D Dirac semimetal Cd_3As_2. Nano Lett., 2022; 22:1138.

Changhua Bao, Haoyuan Zhong, Shaohua Zhou, et al. Ultrafast time- and angle-resolved photoemission spectroscopy with widely tunable probe photon energy of 5.3–7.0 eV for investigating dynamics of three-dimensional materials. Rev. Sci. Instrum., 2022; 93:013902.

Changhua Bao, Laipeng Luo, Hongyun Zhang, et al. Full diagnostics and optimization of time resolution for time- and angle-resolved photoemission spectroscopy. Rev. Sci. Instrum., 2021; 92:033904.

References

1. Nandkishore R, Levitov LS, Chubukov AV (2012) Chiral superconductivity from repulsive interactions in doped graphene. Nat Phys 8:158
2. Hou CY, Chamon C, Mudry C (2007) Electron fractionalization in two-dimensional graphene-like structures. Phys Rev Lett 98:186809
3. Roy B, Herbut IF (2010) Unconventional superconductivity on honeycomb lattice: theory of Kekulé order parameter. Phys Rev B 82:035429
4. Wang WS, Xiang YY, Wang QH et al (2012) Functional renormalization group and variational Monte Carlo studies of the electronic instabilities in graphene near $\frac{1}{4}$ doping. Phys Rev B 85:035414

5. Jiang S, Mesaros A, Ran Y (2014) Chiral spin-density wave, spin-charge-chern liquid, and $d + id$ superconductivity in 1/4-doped correlated electronic systems on the honeycomb lattice. Phys Rev X 4:031040
6. Cayssol J, Dóra B, Simon F et al (2013) Floquet topological insulators. Phys Status Solidi RRL 7(1–2):101–108
7. Bukov M, D'Alessio L, Polkovnikov A (2015) Universal high-frequency behavior of periodically driven systems: from dynamical stabilization to Floquet engineering. Adv Phys 64(2):139–226
8. Wang Y, Claassen M, Pemmaraju CD et al (2018) Theoretical understanding of photon spectroscopies in correlated materials in and out of equilibrium. Nat Rev Mater 3(9):312–323
9. Parameswaran S, Vasseur R (2018) Many-body localization, symmetry and topology. Rep Prog Phys 81(8):082501
10. Oka T, Kitamura S (2019) Floquet engineering of quantum materials. Annu Rev Condens Matter Phys 10:387–408
11. De Giovannini U, Hübener H (2019) Floquet analysis of excitations in materials. J Phys Mater 3(1):012001
12. Rudner MS, Lindner NH (2020) Band structure engineering and non-equilibrium dynamics in Floquet topological insulators. Nat Rev Phys 2(5):229–244
13. de la Torre A, Kennes DM, Claassen M et al (2021) Colloquium: nonthermal pathways to ultrafast control in quantum materials. Rev Mod Phys 93:041002
14. Bao C, Tang P, Sun D et al (2022) Light-induced emergent phenomena in 2D materials and topological materials. Nat Rev Phys 4:33–48